Building a traditional Kurdish City

Hooshmand Alizadeh

Building a traditional Kurdish City

The Urban Morphology of Sanandaj

palgrave
macmillan

Hooshmand Alizadeh
Department of Urban
Planning and Design
Kurdistan Studies Institute
University of Kurdistan
Sanandaj, Iran

ISBN 978-981-16-3633-2 ISBN 978-981-16-3634-9 (eBook)
https://doi.org/10.1007/978-981-16-3634-9

This Palgrave Pivot imprint is published by the registered company Springer Nature Singapore Pte Ltd.
The registered company address is: 152 Beach Road, #21-01/04 Gateway East, Singapore 189721, Singapore

I wish to dedicate my book to the creator of the work of Sharakam Senna

(شاره که م سنه)

Acknowledgements

I would like to take this opportunity to thank those people who over the years supported the successful completion of this book. I start by thanking Najwa Adra (Institute for Social Anthropology—ÖAW), from reading early drafts to giving me helpful advice. Other staff, particularly Hanneke Friedl (Institute for Urban and Regional Research—ÖAW), who helped me out in proofreading three chapters of the book. I would like to acknowledge her help. It is also my pleasure to thank Ali Madanipour (Newcastle University), Stephen Kite (Cardiff University) and Nader Ardalan (Harvard University), who read the entire book and endorsed it by their useful quotes.

And finally, it is to Sanandaj that my gratitude must be directed; to everyone who has helped along the way providing intellectual, physical, or emotional support. I am profoundly grateful to all. I am especially indebted to the late Sirus Marufi and Yadulah Zinoori, to Abdulah Hawari-nasab, Asad Bahrami, Jamal Ahmadi Aeen, Sedeq Munaghi, Asad Gazarani, Abdulhamed Hairatsajadi, Shahoo Keshavarzi. My thanks also extend to my close friends Amir Qaderi and Mansour Subhani, for their outstanding help in preparing information and conducting some parts of the interviews.

Above all, I want to thank my wife, Soheila (Kali), for understanding, encouragement and sacrifice. A lifelong partner makes both the journey and destination worthwhile.

Contents

List of Figures

1

Preamble

Abstract This chapter serves as a preamble wherein an overall view of the book and its constituting parts are discussed including the theoretical basis, the value of traditional settlements, the existing gape in the knowledge of Middle Eastern cities regarding the concept of Kurdish cities, why the city of Sanandaj/Senna, and the structure of the book. It draws attention to the value of the traditional settlements by referring the most important theoretical bases in this regard.

Keywords Traditional urban environment · Kurdish cities · Sanandaj

Introduction

> The reason we are interested in traditional forms of buildings, dwellings and settlements is that we believe that such achievements met human needs in a more sensitive way than contemporary and/or alien methods do. It is this belief that sends us back to the past… (Abu Lughod, 1992 quoted in Bianca, 2000, p. 184)

H. Alizadeh, *Building a traditional Kurdish City*,
https://doi.org/10.1007/978-981-16-3634-9_1

Urban and architectural artefacts of the traditional environment were shaped by human intention and intervention and can be seen as important imprints of human beings which have evolved over centuries according to strong socio-cultural and environmental "rationalities" through which we can extract the priorities and beliefs of the decision-makers (Ergecgil, 2001). That is why "the lessons that are to be found there are not so casually learned", Paul Oliver said (Oliver et al., 1981: 204). This means that traditional cities' morphology and spatial patterns serve as the content of meaning:[1] people live where they can have a meaningful experience of the environment. These morphologies and spatial patterns have gradually evolved to meet their populations' cultural needs and, at the same time, to respond to their surrounding environment (Alizadeh & Irandoost, 2017; Norberg-Schulz, 1980). Therefore, the traditional urban environment is considered here as the materialised form of human thoughts, skills and resources which created "functional, meaningful and identifiable spaces in relation to society, time and place" (Ozaslan, 1996: 3).

Thus, traditional urban environments can be considered irreplaceable assets representing the investment of centuries of knowledge and lessons which weave the various physical elements and spatial units together in particular contexts. For this reason, such environments are "a treasure house of human experience – of successes and failure" (Rapoport, 1989: 100), good examples of "problem solving a total response to the challenge of climate and topography" and demonstrations of human adaptations to the ebb and flow of social and political history of particular nations (Ahmad & Malcolm, 2001: 72). Generally speaking, these environments can be considered laboratories to probe the process of urban environment formation and the interrelationship between socio-cultural and physical environments.

Scholarly publications on the urban form of Kurdish cities are rare, and their distinctive characteristics are largely unknown to scholars of Middle Eastern urban spaces. Studies of Iranian cities have paid little attention to Kurdish cities. Most Iranian literature on the traditional

[1] The meaning of any object consists in its relationships to other objects, that is, it consists in what the object gathers (Norberg-Schulz, 1980: 166).

built environment focuses on the Iranian plateau, especially cities located in the centre of the plateau and its surrounding area on the slopes of mountain ranges in the hot-arid zone (Bonine, 1979; English, 1966; Habibi, 1996; Kiani, 1986, 1987; Kheirabadi, 1991; Lockhart, 1939; Tavassoli, 1982). No scholars have studied Kurdish cities located in the west and north-west of Iran in relation to Kurdish habits and traditions. Clark and Clark (1969), in their study of the city of *Kermanshah*, one of the largest Kurdish cities in the western part of Iran, explored the process of city development over the course time using a sequence of aerial maps without reference to the role of Kurdish culture in the shape and organisation of the various parts of the city. The only exception is the work of Vasilyeva (2000) who, for the first time, emphasised the concept of the Kurdish city in her study of *Sanandaj*, and some works of the author published in peer-reviewed journals. She gathered valuable information concerning the historical background of the city. But her work does not deal with spatial dimension of the city and lacks comparative discussion with other Iranian cities to draw out their distinctive character taking into account the role of Kurdish culture. This book will fill an important gap in the knowledge of Middle Eastern cities by reconstructing the historical process of Sanandaj's formation and development until the rise of modernisation in Iran at the end of Qajar dynasty (1785–1925). An interpretive-historical approach will be applied to explain how urban form in a particular context responded to the environmental and socio-political determinants of the time. This will also shed light on the features of the built form as they relate to Kurdish customs and traditions. The book will contribute to an understanding of the nature of Kurdish settlements by paying attention to the interaction between the multifarious social and spatial forces that have conditioned the processes and patterns of city formation and development over time. In line with this, it will point to similarities and differences with other Islamic and Iranian cities. This comprehensive insight into the morphology of Sanandaj responds to the clear need to consolidate the scarce formal literature and local documentation on the city. In addition, it calls for a methodological approach in which the investigation of any individual example should aim at establishing clearly what is unique and peculiar to the place and explicable only through understanding local conditions.

More importantly, this book can also provide a foundation for further studies of other Kurdish cities in the region particularly in Iraq, Turkey and Syria where most Kurdish cities are located.

The primary objective of this book is to assist those who are interested to understand the urban form of Kurdish principal cities. It will be an excellent resource for students and researchers of urban studies, geography, social science and Kurdish studies.

Why the City of Sanandaj/Senna?

Sanandaj is the capital of Kurdistan province in north-western Iran, with a majority Kurdish population. It acquired its peculiar character from its well-known location within the Zagros Mountain range—the mountains that define the western edge of the Iranian plateau—and to Kurdish cultures. The Zagros mountains have been recognised as the physical phenomenon with the greatest influence in the shaping of Kurdish history, people, traditions and culture (Gunter, 2018; Izady, 1992). Historically, the mountains provided a majestic border between two rival empires: the Ottoman Empire (1299–1922) which adhered to the Hanafi (Sunni) tradition and the Safavid Empire (1501–1722) which was Shiite. These mountains played a major role in the formation of independent castellated city structures, often located in inaccessible places during the Ardalan Dynasty, a Kurdish dynasty that ruled north-western Persia (Iran) from 1169 until 1867. In contrast to these rival empires, the majority of Kurdish Muslims adhered to the *Shafi'ite* Sunni rites in order to distinguish their identity from the surrounding lands.

The nature of the challenges between those rival empires and the independent status of castellated city structure of Ardalan's dynasty in the "Perso-Turkish" frontiers (Minorsky, 1943) provoked the rulers of these two empires to terminate this structure in the treaty of Zahab in 1639.[2] As a result of this treaty, four castellated cities of Ardalan's region were

[2] The Treaty of Zuhab, also called Treaty of Qasr-e Shirin, was an accord signed between the Safavid Empire and the Ottoman Empire on 17 May 1639. The treaty confirmed the dividing of territories between these two rival empires.

destroyed and a new city of Senaa-Dej (Sanandaj) to some extent capable of control by the Iranian side was created. The city of Sanandaj was founded in 1639 by the Ardalan family with an unusual form distinct from other Iranian cities. Its structure shows a particular character that differs from other traditional Iranian cities had shown before the start of modernisation in Iran. Its main structure sprang from castellated cities, which contained three layers of urban space: the citadel or Qala on the top of a hill in the centre of the city was surrounded by a wall, the quarter occupied by the ruling class was also enclosed within a wall and, beyond these two parts, the rest of the city comprised three main neighbourhoods and one newer free of man-made walls (Figs. 2.4 and 2.8). These three quarters resembled a collection of villages circling the nucleus of power. They were formed and organised primarily based on the professions. These professions were in line with the needs of ruling class and a medium-sized city function. This city is, therefore, an important built form which presents Kurdish understanding of the environment.

It is also important to emphasise that the city's foundation and development were bound to the destiny of the *Ardalan* family, who left its major imprints on the city's structure. To some extent, when this family was driven out of the government of Kurdistan in 1867 the city also lost its splendid image. "As the symbol of the *Ardalans*' power *Sanandaj* followed all the curves of their political fate" including the continuous efforts of this family to rebuild and redecorate its constructions (Vasilyeva, 2000: 5). The change of government to non-local rulers ended this city's unique atmosphere informed by Kurdish customs and traditions. More importantly, the new rulers engaged primarily in military action to pacify the tribes on the frontier and quell popular protests against their rule. For all the above reasons, the traditional Kurdish city of Sanandaj is a unique case worthy of study and publication to reconstruct the historical process of formation and development until the rise of modernisation in Iran at the end of Qajar dynasty (1785–1925).

The Structure of the Book

This book is structured in seven chapters. This chapter serves as a preamble wherein an overall view of the book and its constituting parts are discussed. Chapter 2 follows Lambton's (2011) assertion concerning the importance of each city's own history. It provides a chronological account of the city up to the first two decades of the twentieth century. Based on historical records, this chapter will address two main periods—before and after the Safavid period (1501–1736). The Safavid dynasty, especially the period of Shah Safi (1629–1642) covers the time of the city's foundation. Although, this period does not relate directly to the city, it aims to provide a clear contextual picture of the region and the factors that resulted in the city's foundation and its subsequent spatial organisation. The second stage extends from 1639 to the Pahlavi dynasty, the period of modernisation which led to huge socio-economic and physical changes in Iranian cities (1925–1979). These materials cover the origins of Sanandaj, the emergence of a Kurdish town and narrates the process of Sanandaj's formation and changes in the course of time. The resulting narrative will elucidate the process of city formation and development through time. It will explain changes that have occurred and the socio-political movements and dynamics that gave the city its particular distinguishing character. The aim is to reveal the evolutionary process of city building and its main shaping factors. This historical material is presented in a concise and illuminating manner through the selective use of photographs, maps and diagrams.

Chapter 3 examines the city's morphological components through contextual analysis in two sections that together make clear the socio-spatial rules underlying the city's morphological components. In the first section, the city's built form within its macro and microgeographical settings is examined to reveal the nature of the interaction between man-made and natural environments and to identify the features of the city that responded to the natural landscape. This section builds on the concept of "being-in-the-world" developed by Heidegger. For him, the idea of dwelling is conceptualised from a relational perspective, particularly the way we exist in the world/space and the way we make the world/space meaningful, or place-like (Cresswell, 2014; Shah, 2015). It is the way a unique gathering of things, human meanings and values endow and transform lived space (Gieryn, 2000; Tuan, 1977). The second section of this chapter examines the city's constituent elements, especially the morphological elements of social communication and spatial articulation. It provides a dynamic perspective, revealing the city's fixed points, details and the logic of urban configuration to identify those that have modelled the urban fabric. City walls, residential quarters (*mahallaha*) and public spaces will be discussed.

After the morphological features of the whole city's traditional part, the following chapters provide more details of socio-spatial structure of Sanandaj's traditional built form by focusing on one of the neighbourhood, the built form of which has been changed less than others and is consequently more appropriate to understand the unfolding of the socio-spatial principles underlying the constituent elements of the traditional built form. In this part, Chapter 4 examines structural elements of the Qatar-chyan neighbourhood in detail, including the features of the terrain, the socio-spatial factors that gave birth to the neigbourhood, influenced the development and use of urban space and eventually led the neigbourhood to acquire a character distinguishable from other neighbourhoods of the city. This chapter elaborates Hall and Barrett's assertion that built form is "the outcome of a complex mix of socio-economic and cultural forces and the actions and intentions of a variety of actors" (2012: 30). Urban form does not "simply appear overnight, like a movie-set springing up on a vacant lot, but has to be produced" (Alizadeh & Habibi, 2008: 153). The point is also emphasised by Rossi (1982) who argues that studying the built environment does not imply only the visible image of the city and the sum of its different parts. The city must be viewed as a totality which has been shaped over time by human activity (see also Kostof, 1991). This was recognised as the most comprehensive way of analysing the city because it implies the process of city construction as a human creation (Kostof, 1991). Thus, the first point in dealing with such an environment is considered to be the tracing of time in the evaluation of the city and keeping in mind that the whole is more than its parts. This point becomes naturally understood when the process of city building is probed; because the notion of process in itself comprises the relatedness among the socio-cultural events that resulted in urban artefacts. While the role of nature as form giver is an important factor, we cannot undermine other soft factors. This also means considering a large number of factors, especially socio-cultural ones, involved in this process. By doing this, we can relate the physical environment with the social, "relating the world of artefacts with the world of people"—who built and used them (Alizadeh & Habibi, 2008: 153). This defines the nature of built form which comprises both the realms of the physical and the social and multidimensions background namely as socio-spatial factors.

"As cities of all cultures and the social life of their citizens, at all historical periods, are organized along some form of public-private lines" (Madanipour, 2003: 1), after analysing the structural elements of the quarter, Chapters 5 and 6 discuss the two domains of public and private spaces encompassing the social life of the neighbourhoods. These will be examined from both social and physical dimensions to approach the design principles that aggregated these realms and also the ideas and beliefs which contextualised them in that society. The way space is subdivided and the relationship between the public and private spheres in general is a mirror of social relations and a main indicator of how a society organises itself. Alongside these issues, some considerations on the experiential qualities of physical space in the circulation system of the neighbourhood are discussed.

As most of the literature on Islamic cities deal predominantly with the male's domain, Chapter 5 highlights the women's domain, and its main points explore the role of women in the social life of the quarter within the Kurdish city of Sanandaj. To clarify the subject, a few questions are addressed in this chapter as follows: do women relate to the public sphere in ways other than fear? How do men relate to domestic space? What are gender relations within the household? Do Kurdish public/private spaces differ from those in the rest of Iran? A comparative perspective with Iranian and Islamic contexts is adopted to make sense of the features of public life in this context. The chapter delineates the intricacies of how Kurds conceptualised the public domain, and specifically how females have experienced that domain, with regard to the common view of "unsafeness". We also describe the role of women and their relationship to space outside of kinship ties. Considering the structure of the book which starts from the whole to the parts, first, the discussion deals with public spaces of the selected neighbourhood as the main domain of social life, and then it moves into the private spaces of the houses. Between the two realms, I will analyse the cluster of houses to discover the manner of aggregation, neighbourhood structure and the transition to the private domain. The private domain itself is discussed in Chapter 6. External and internal order are interdependent. The way each develops depends on what happens in the other and vice versa (Rapoport, 1977). Thus, this chapter investigates the private domain to grasp the continuity of the order perceived in the public spaces with the private spaces, especially the courtyards of houses. In this journey, I will analyse the spatial connections to the outside world and the spatial organisation of the houses into two realms of higher and lower status to highlight the different values and attitudes attributed to them. This comparison will also show the distinct character of the housing settlements and their differences with other Islamic cities, especially Iranian ones. This is important because the houses of notables are to some degree built on the model of other Iranian cities due to owners' socio-political relationships with those cities, while the common dwellings reveal more indigenous patterns.

The last part of the book comprises Chapter 7 which outlines the book conclusions and findings with some recommendations for future studies. It also draws attention to some suggestions concerning the findings and their application in future prospects of the city and similar situations.

References

Ahmad, K., & Malcolm, P. (2001). Traditional settlements, user participation and 'no harm' principle toward a contemporary participatory process in low-income public housing in Islamic cities. In H. Turgut & P. Kellett (Eds.), *Traditional environments in a new millennium* (pp. 69–74). Istanbul Technology University.

Alizadeh, H., & Irandoost, K. (2017). Investigating the process of traditional design principles formation in the Iranian-Kurdish urban quarters. *Urban Design International, 22*(3), 197–218.

Alizadeh, H., & Habibi, K. (2008). Structural elements and the built environment. *International Journal of Environmental Research, 2*(2).
Bianca, S. (2000). *Urban form in the Arab world: Past and present*. Vdf.
Bonine, M. E. (1979). The morphogenesis of Iranian cities. *Annals of the Association of American Geographers, 69*(2), 208–224.
Clark, J. I., & Clark, B. D. (1969). *Kermanshah: An Iranian provincial city*. University of Durham.
Cresswell, T. (2014). *Place: An introduction*. Wiley.
English, P. W. (1966). *City and village in Iran: Settlement and economy in the Kirman Basin*. University of Wisconsin Press.
Ergecgil, S. C. (2001). Modern culture and its perception of tradition, history and historical sites. In H. Turgut & P. Kellett (Eds.), *Traditional environments in a new millennium* (pp. 75–78). Istanbul Technology University.
Gieryn, T. F. (2000). A space for place in sociology. *Annual Review of Sociology, 26*, 463–496.
Gunter, M. M. (2018). *Historical dictionary of the Kurds* (Vol. 8). Rowman & Littlefield.
Habibi, S. M. (1996). *Az Shar Ta Shahr* (Analytical review of the city concept and its physical image in the course of time). University of Tehran.
Hall, T., & Barrett, H. (2012). *Urban geography*. Routledge.
Izady, M. R. (1992). *A concise handbook: The Kurds*. Crane Russak (Taylor & Francis).
Kheirabadi, M. (1991). *Iranian cities: Formation and development*. University of Texas Press.
Kiani, M. Y., (Ed.). (1986). *Shahr-ha-yi Iran* (Iranian cities). Jahad-e Danishgahi.
Kiani, M. Y., (Ed.). (1987). *Nazari Ijmali bih Shahr-nishini va Shahrsazi dar Iran* (A general study on urbanisation and urban planning in Iran). Farhang va Ershad-e Eslami.
Kostof, S. (1991). *The city shaped: Urban patterns and meanings through history*. Thames & Hudson.
Lambton, A. S. (2011). Administration and social organisation. In E. Yarshater (Ed.), *Encyclopaedia Iranica: Cities*. Columbia University and Bibliotheca Persia.
Lockhart, L. (1939). *Famous cities of Iran*. W. Pearce & Co.
Madanipour, A. R. (2003). *Public and private spaces of the city*. Routledge.
Minorsky, V. (1943). Sanandaj. In M. T. Houtsma, A. J. Wensinck, H. A. R. Gibb, W. Heffening & E. Levi-Provencal (Eds.), *Encyclopaedia of Islam*. E. J. Brill and Leyden & Luzac & Co.

Norberg-Schulz, C. (1980). *Genius Loci: Towards a phenomenology of architecture*. Rizzoli International Publications.
Oliver, P., Davis, I., & Bentley, I. (1981). *Dunroamin: Thesuburban semi and its enemies*. Barrie & Jenkins.
Ozaslan, N. (1996). *Historic urban fabric: Source of inspiration for contemporary city form* (p. 373) (Dept. of Architecture. Unpublished Ph.D). University of York.
Rapoport, A. (1977). *Human aspects of urban form: Towards a man-environment approach to urban form and design*. Pergamon Press.
Rapoport, A. (1989). On the attributes of tradition. In J.-P. Bourdier & N. Alsayyad (Eds.), *Dwellings, settlements and tradition: Cross-cultural perspectives* (pp. 77–105). University Press of America .
Rossi, A. (1982). *Architettura della città*. MIT Press.
Shah, R. (2015). *Making sense of disability: The role of home and its physical environment* (p. 261) (Doctor of Philosophy). School of Design. Creative Industries Faculty, Queensland University of Technology.
Tavassoli, M. (1982). *Sakht-i Shahr va Miamari dar Iqlim-i Garam va Khushk-i Iran* (Urban structure and architecture in the Hot-arid Zone of Iran). Urban Planning and Architecture Research Centre of Iran.
Tuan, Y.-F. (1977). *Space and place: The perspective of experience*. University of Minnesota Press.
Vasilyeva, EI. (2000) The Kurdish city of Sanandaj: Ardelan's Capital. Paris, conference presentation.

2

Historical Background

Abstract This chapter provides a chronological account of the city up to the first two decades of twentieth century. It covers the origins of Sanandaj, the emergence of a Kurdish town and narrates the process of Sanandaj's formation and changes in the course of time. The resulting narrative will elucidate the process of city formation and development through time. It will explain changes that have occurred and the socio-political movements and dynamics that gave the city its particular distinguishing character. It reveals the evolutionary process of city building and its main shaping factors. This historical material is presented in a concise and illuminating manner through the selective use of photographs, maps and diagrams. It was found that the main reason of its building lies in the essence of political challenge between two rival empires in the seventeenth century, Ottoman and Safavid empires on opposing sides of the Zagros Mountains. It was also became clear that the city's destiny was completely bound with the power of Ardalan's family in the course of time.

Keywords Historical background · Sanandaj/Senna · Mound-cities · Ardalan · Mobile capitals · Castellated cities

H. Alizadeh, *Building a traditional Kurdish City*,
https://doi.org/10.1007/978-981-16-3634-9_2

Introduction

> Each city has its own story to tell, but not all are recorded, and what is true of one city at one period is not necessarily true of all periods or of other cities. (Lambton, 2011)

We now provide a chronological account of the history of the city until the first two decades of the twentieth century. This includes two main periods—before and after the Safavid period (1501–1736). The city was founded during the Safavid dynasty, especially the period of Shah Safi (1629–1642). Although the period prior to the Safavid dynasty does not relate directly to the city, we aim to provide a clear contextual picture of the region and of the factors that influenced the city's foundation and its subsequent spatial organisation. The second stage extends from 1639 to the Pahlavi dynasty (1925–1979), the period of modernisation which led to huge socio-economic and physical changes in Iranian cities. Considering the historical evidences and an analysis of the existing built form of the city, most of the discussion is devoted to the second period, which contains many sub-periods and during which the evolution of the whole city took place.

This section covers the origins of Sanandaj and narrates the process of its formation and changes in the course of time. The resulting narrative will elucidate the process of city formation and development through time. It will explain changes that have occurred and the socio-political movements and dynamics that gave the city its particular character. The aim is to reveal the evolutionary process of city building and its main shaping factors. This historical material is presented in a concise and illuminating manner through the selective use of photographs, maps and diagrams.

The primary source of historical information in urban studies was recognised in line with existing buildings and spaces which contain many clues about the evolution of urban and social environments (Saleh, 1995). But as Kostof (1991: 10) indicates, in order to derive those clues, especially socio-political ones, the physical form of the city itself does not tell us "what really happened until we turn to the archives, the history books, the old maps – until we assemble all the evidence, some of it often contradictory, that will help explain how a particular downtown got the look it now has". However, given the lack of a comprehensive source of documents and written history of Sanadaj, first-hand information was obtained from field documentation and partly from oral history.

Etimology of the Name of the City

Sanandaj is pronounced in Kurdish as "Senna" or "Kursan". There are different viewpoints concerning the etymology of Sanandaj. For Ali Akbar-e VaqaeaNegar, Sanandaj was originally only one more village among Kurdistan's villages and was known as "Seenna". Over the course of time, "Seenna" came to be pronounced as "Senna", and due to the castle that was built on top of a hill, with Seenna located on its slope, it came to be called "SennaDej" (the castle of Senna). Later, this presumably turned into Sanandaj (Kurdistani, 2002). According to Mardukh, "Dej" or "Dez" means castle in Kurdish and Old Persian. Therefore, "Senna Dej" signifies the castle of Senna, or the capital of Senna. Over time, the words have become Arabic, written as سنه دج.. And finally, as a result of such Arabisation and distortion, the term "Sanandaj" was formed and used to name the city (Mardukh, 1972). From the viewpoint of Kamandi, the castle that was built during the reign of Toos-nauzar[1] over a hill with the same name in the southeastern part of the current city of Sanandaj was a gathering place for Khans and leaders of the surrounding regions. For this reason, "Sunan Dej" became known as "Khans' Fortress". This etymology goes back to the dialect of Orami in the Kurdish Language because the word "Sun" means "Khan" in Orami (Kamandi, 2014). For Shirin Ardalan, Sanandaj's original name refers to

[1] The commander in chief, Toos-nauzar is one of those afflicted princes in the ancient book who had served prominent Kayanid kings such as Kay Kavus and Kay Khosrow. Tus is of the great clan of Nauzar and is an eternal character in Zoroastrian religion. On the day of resurrection, he comes to assist the Saoshyants and it is he who starts the apocalypse (Vaceq Abbasi and ABDI, 2017).

the legacy of Yarsani in the region (Ardalan, 2008). The name of "Senna-Sanandaj", the capital of the Ardalan dynasty,[2] is probably derived from the word "Sunan Dej", which means the fortress of Sunan (Sultans—the embodiment of the universal spirit). For others, the name of Sanandaj came from the two words Seinna and Dej. There was a small village on a hillside (Seinna) onto which a castle (Dej) was built by the Wali Sulayman Khan on the ruined site of an earlier settlement. For this reason, the city was named "Seinna-dej" and over time it changed, due to various pronunciations, until the first period of the Pahlavi dynasty, when it became, as it is now pronounced, Sanandaj.

The Period Prior to the Foundation of the City

The period prior to the foundation of the city can be classified into two sub-periods: the pre-Islamic and the period during the caliphate until the Mongol invasion (1219–1221). For the pre-Islamic period, the local and other historical records do not show any clear trace of a city in the current location of Sanandaj. The presence of sacred sites within the city's area and its geophysical borders however imply that the city setting was presumably an arena for human settlements in this period, especially during the Sasanid Empire (224–641 A.D.). These include the mound of Toos-nauzar in the eastern part of the city, a place commemorating a great Iranian champion that lay like sleeping turtles upon the landscape, and the Awear Mountain in the south as "the second place of celestial revelation for Zoroaster"[3](Dultshahi, 1984: 88, 150) (Fig. 2.1). "People usually kept the old Quran between the rocks on the top of the mountain in Awear" and named it the Kuchick Quran, the stone of the Quran (Alizadeh et al., 2019: 6).

[2] The founder of Ardalan's genealogy was Baba Ardalan, who was known as a descendant of Ardashir Babakan (the founder of the Sassanid dynasty in 226 A.D) (Ardalan, 2008: 4) and who reconstructed the fortresses (four Castellated towns) in order to extend and maintain the power of his family over the region of Kurdistan.

[3] Zoroastrianism is the oldest of the revealed creedal religions, and it has probably had more influence on humanity, directly and indirectly, than any other single faith (Boyce, 2001:1–2). It was the dominant religion in Iran until the triumph of Islam.

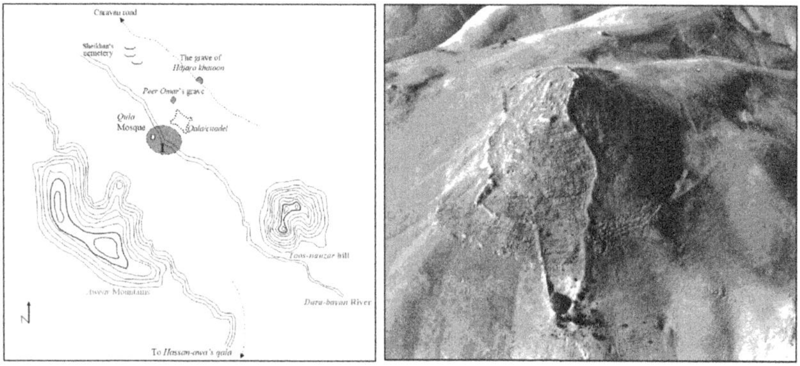

Fig. 2.1 Left: The main features of the city site before its foundation. Right: oblique aerial view of the remnants of Hassanabad Qala in the period of mound-cities which shows its planned structure (*Source* Left drawing by author, Autumn 2005, right image from Schmidt, 1940)

In supporting the above notion concerning the mound of Toos-nauzar, Mardukh (2000: 263) says, the main residents of Senna Dej (Sanandaj) were the clan of Zarien-kawsh (golden shoes), who were the followers of Toos-nauzar. The foothold of the famous Toos-nauzar was the hill down from the city, which separates the city from the river of Qeshlaq. This hill became known as the hill of Toos-nauzar and some of his followers still live in Sanandaj (Mardukh, 2000).

For the period of the Islamic Empire onwards, especially the period after the Caliphate, there are some points in the literature (Kurdistani, 2002; Mardukh, 2000; Sanandaji, 1996; Soltani, 2006) and even some evidence on site, confirming the presence of settlement on the site of the city. For example, for Kurdistani

> Senna was built by the Wali Sulayman Khan on the ruined site of an earlier settlement. (2002: 27)

Apart from this, the cemetery of Sheikhan, located in the west of the city (Fig. 2.1), which contained old graves of masheikh (plural of sheikh) dating back nine hundred years. For more evidence, it is possible to draw attention to the graves of Hajara Khatoon and Peer-Omar. The first was identified by local historians as the sister of Imam Reza (the eighth Imam of the Shiite sect). It was located alongside the course of the city's wall close to the northern gate and the probable main caravan road to the west of Kurdistan and the Iraq region (Fig. 2.1). In the case of the second grave, there are differences among the historians: some identify it as the grave of Imam Ali's son (the first Shiite Imam and the fourth Sunni Caliph) and some others ascribe it to Imam Ali Zain al-Abedin, the fourth Shiite Imam. Its location is close to the seat of power, the Qala/citadel (Fig. 2.1), adjacent to the Arg square in front of the Qala. The presence of these graves highlights the site of Sanandaj as an important place during the Islamic Empire, because the imams and their descendants usually settled in and visited important and populated regions in order to carry out their duties in guiding the people towards the religion of Islam.

In line with the above evidence, as witnessed by some historians (Bayat, 1988; Minorsky, 1943; Nikitine, 1956), there was a main centre on the site of Sanandaj, or close by, named Sisar (thirty heads). In the words of Le Strange, this account can be traced as follows:

> About sixty miles north of the ruins of Dinavar stands at the present day the important town of Sihnah (Senna), which is the modern capital of the Persian province of Kurdistan, though under this name it is not mentioned by any medieval Arab or Persian geographers. In the position of the modern Sihnah, however, according to the itineraries of Ibn Khurdadbih and Kudamah stood, during the middle-ages, the city of Sisar, a name which Yakut righty says means in Persian "thirty Heads". The nieghbourhood of Sisar abounded in springs and was known as the Sad-Khaniya – the Hundred Houses or Heads of Water – from the number of these springs. The Caliph Amin had built a fortress here, which his more celebrated brother Mamun had garrisoned, taking into his pay the Kurdish tribes who held the sourding pastures, and using them in the civil war against his brother, whom he deprived later on of the Caliphate. Sisar was counted that the modern name of Sihnah may be merely a corruption of Sad-Khaniyah, shortened to Si-Khanah, thirty Houses, but of this is no direct evidence. (Le Strange, 1905: 190)

That is why Vasilyeva (2000: 1) ascribes the foundation of Sisar to the period of the Caliphate as a centre to "keep a military contingent in order to maintain the contacts with Kurds". As Shapur Shahbazi and Edmund Bosworth (1990) point out, "the Arab warriors who overthrew the Sasanians normally established garrisons in existing urban centers and in many cases adjacent to the frontier war zones" to control the captured areas. Considering this evidence, it can be claimed that the site of Sanandaj was populated before its foundation by the time of the Mongol invasion (thirteenth century), during which a new era of mobile capital started in the region.

The Period of Mound-Cities

The intrusion of the Mongols in the thirteenth century, who were accustomed to a high degree of mobility and who were suspicious of fixed residences, caused the collapse of the city's organisation and more generally, the decentralisation of power within Iran. This means that the nomadic way of life dominated over the urbanity in Iran leading to the demolition of many cities throughout Iran, and this formed a new approach to life named "defence-plunder" (Habibi, 1996). In addition to the city of Sisar there were a number of other flourishing cities in the region, for instance Shahrizour, Abhar and Suhravard, which were destroyed by the Mongols (Vasilyeva, 2000). These invasions and, later on, the division of the Middle East into two warring empires (Persian and Ottoman), with their line of fire being the heartland of Kurdistan (Izady, 1992), resulted in devastating wars in later periods. The lack of security and "the disruption of the overland east-west trade" led to the formation of another type of settlement in the region as human settlements were, in general, at high risk of attack and being plundered (Izady, 1992: 121). The surviving inhabitants took refuge in strongly fortified and shrunken urban islands on the tops of the high mountains in order to defend themselves. This movement recalled the old concepts of city organisation in the form of mound-towns or Qala-towns (castle-like towns) built in inaccessible places (Izady, 1992). The root of this can be traced back to the period of the Median Empire (728–550 BC), when the organisation of the country was based on the mound-towns located in strategic places. It usually comprised a strong Qala (castle) on the top of a mound which implied a planned structure (Fig. 2.1). The settlements were usually located on the slope of the mound and behind the battlement walls of the Qala (Habibi, 1996).

As Vasilyeva states (2000: 3), the formation of such settlements shows that "the history pendulum moved toward the pastoral morpheme, tribal traditions and nomad settlements" (Vasilyeva, 2000: 3). Due to such circumstances in the region, it appears that the site of Sanandaj became depopulated or disregarded by the rulers as a seat of power for several centuries; instead, people lived on the top of inaccessible mounds and places in the Zagros mountain range (Fig. 2.1). For the same reason, the literature shows no trace of settlement as a town on the site of Sanandaj from the Mongol period up to the foundation of the city in 1639. The only exception is the explanation of local historians about the etymology of Sanandaj's name. Sanandaji (1996), Mardukh (2000) and Kurdistani (2002) explain that when Wali Sulayman Khan was charged to change the site of power to another place by Shah Safi, he selected the site of Sanandaj, where there was a village named Seenna. As he created or reconstructed the Qala which was near to this village, it was known Seenna's Qala (Senna-dej). Continuous use of these two words merged them together as a one word—Sanandaj. Although the return to defensive town appeared to be a step backwards, it led to a kind of semi-independent government structure on the "Perso-Turkish" frontiers (Minorsky, 1943) due to the inaccessible situations of the mound-towns (particularly the capital towns) in the Zagros Mountains. It also gave rise to the Ardalan family as rulers of Kurdistan for more than seven centuries (116–81,867) (Izady, 1992). During this period (ranging from the Mongol invasion to the city's foundation in 1639), the structure of mound-towns including four castellated towns as "mobile capitals" dominated over the region of Kurdistan by the rulers of Ardalan. On the importance of those Qala, Mastoureh Ardalan[4] (quoted in Sharafkandi, 2002: 62), who wrote the history of Ardalan and lived in 1805–1848, said:

[4] She was the first Kurdish-Iranian female historian. She studied Kurdish, Arabic and Persian literature under the supervision of her farther, Abolhassan Beig Qadiri (Mizanian, 2017).

> I have seen that the ambassadors of Russia and England visited Kurdistan, in the period of Amanulah-Khan the great (1800-1825), in order to view the greatness of the Qala's remains. (Hasanabad Qala)

As Vasilyeva (2000: 3) explains, "the inhabitants of Hasanabad Qala were numbered two thousand families or at least twelve thousand people that is equivalent of the population of a medium size city" in the period of Khan Ahmad-Khan (1616–1636).

To understand the above castles and their locations, historical records (Babani, 1998; Kurdistani, 2002; Mardukh, 2000; Vasilyeva, 2000), reveal that under the leadership of Kulul (1258–1271), the son of Baba Ardalan, other Qalas were reconstructed as he changed the site of his power from the castle of Zalm in the Iraqi region, far to the west of Sanandaj, to Palangan, located far to the south of Sanandaj. In this location, he also expanded his power and the territory of Ardalan to other regions by reconstructing two other Qalas at Marivan in the west and Hasanabad to the south of Sanandaj. The locations of these Qalas comprised the whole territory of Ardalan, which stretches across the mountain range of Zagros as shown in Fig. 2.2.

As the local chronicles explain, each Qala contained the main elements of a traditional Iranian town including the bazaar, mosque and hammam (bath house) in the absence of an active central city in the region. They were well defined by nature and man-made structures as castellated towns on top of inaccessible mounds. This manner of settling in the natural landscape was continued by other rulers of Ardalan's family up to the period of Shah Safi in the first half of the seventeenth century. The important point is the way the Qalas were used by the Ardalan rulers. As local historians indicate, Halo-Khan (1590–1616) used the Qalas as "mobile capitals", based on the seasonal change in order to sustain and strengthen their power over the whole region of Kurdistan. For example, Mardukh (1972, quoted in Sharafkandi, 2002: 66) clarifies that Halo-Khan used these Qalas for himself as follows: Zalm as a winter qala, Marivan during the autumn, Palangan during the spring and Hasanabad as summer Qala and the main centre for Ardalans' government close to the current location of Sanandaj (five kilometres away).

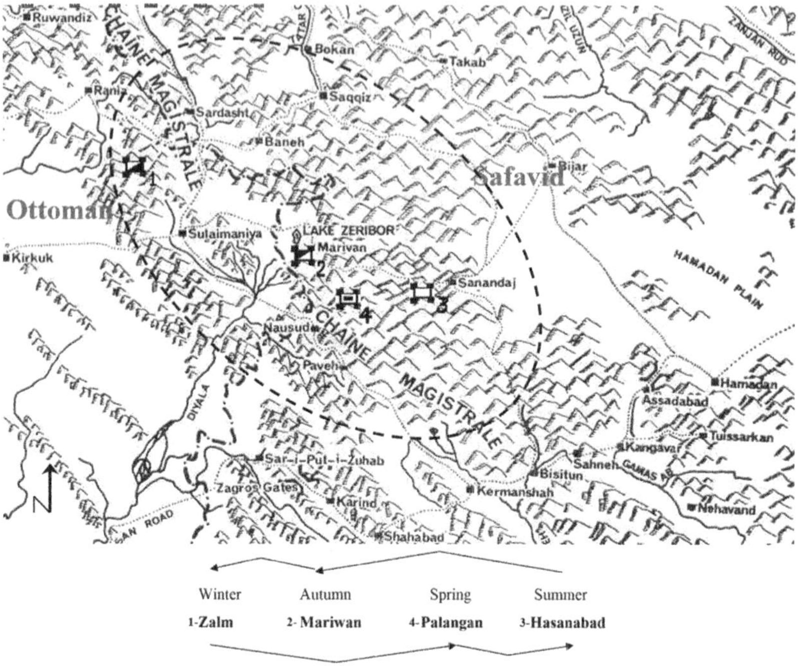

Fig. 2.2 The locations of four Qalas in the territory of Ardalan in the frontiers of two empires (Ottoman and Safavid) (The Qalas of Hasanabad and Zalm served as the points of departure and arrival, as the former is in the south-east and the latter in the north-west of Ardalan territory: map from Levine [1974: 101] with some adding points by author, Autumn 2005)

Presumably the climatic conditions of the region as well as political considerations influenced such mobile use of capitals. The good spring weather in Palangan combined with the fascinating natural landscape appointed it as the spring capital, and Zalm with its modest climate as the winter Qala. Consequently, in order to control the whole territory of Ardalan, the other Qalas were appointed to be used during the autumn and summer. From the point of political considerations, using mobile capitals strengthened the power of the Wali of Ardalan in controlling the whole region and enjoying an independent status in the "Perso-Turkish" frontiers (Minorsky, 1943).

As Wali Halo-Khan and Shah Abbas the Great (1587–1629) were contemporaries, the mobile capitals and independent status of Wali was intolerable to Shah Abbas, who ensnared him with a political trick. As Safavid leaders were aware of the long-term stability of Ardalan's Khans and their great power in the region, they tried to engage them into their government through intermarriage policy, later on assigning to them the land and ceding some parts of the state. This was a certain attempt towards "Iranicising" the Ardalans' family. In this regard, the Shah compromised with Wali through intermarriage and Khan Ahmad-Khan, Wali's son, was married to the Shah's sister. Apart from conquering the territory of Ardalan, the main motive behind this decision was to further the Iranicising of the Ardalan family and attract them into the land of Iran rather than the land of the Ottoman Empire on other side of the Zagros Mountains. Other convincing evidence comes from Safrastian (1948: 42) who states that:

> Shah Abbas commissioned Prince Sharafedin, the famous Kurdish historian and a scion of the Sharaf Khan Dynasty of Bitlis, to write, in Persian, a history of the Kurdish tribes and their genealogies. His aim was, of course, to arouse the hereditary pride of the Kurds to recall their ancient associations with Persia and the east, and thus to divert their attention from the Ottoman Empire.

As a result of the above political decision, for the first time in the political life of the Ardalan dynasty the power was transferred to Wali's son (Khan Ahmad-Khan, the Shah Abbas's son-in-law) by the order of the Shah during the lifetime of Wali. From one perspective, this political movement of Shah Abbas can be recognised as a starting point in changing the site of power from mound-towns to relatively level areas in the current location of Sanandaj. In line with this notion, Ahmadi[5] and Marufi[6] claim that the idea of changing the seat of power was first introduced to Khan Ahmad-Khan because he was influenced by the view of his wife. As they explain, the reason behind this claim can be attributed to the way of life of Wali's wife, who experienced the civic life of Isfahan

[5] Interview 3 July 2003.

[6] Interview 27 August 2003.

as a princess in the court of the Shah. From this point, they conceived that the garden ensemble of Bag-e Meidan (located on north side of the traditional city which later on composed the quarter of Char-baq) was first structured by Khan Ahmad-Khan in favour of his wife. Vasilyeva (2000: 4) also points out that

> The garden ensemble of Bag-e Meidan was created from the remnants which existed from the time preceding the valis of Ardalan.

But as the challenge of power did not give Khan Ahmad-Khan enough opportunity, he could not pursue the idea of changing the seat of power to the site of Sanandaj and it was carried out by his successor who was appointed by the Court of the Safavid Empire.

For Vasilyeva (2000: 3), the main inducement to return to the city is the conclusion of the Iran–Turkey peace treaty of Zuhab contracted on 17 May 1639—"they decided to destroy all Kurdish citadels". Under the act of this peace treaty, Sulaiman-Khan Ardalan (who was one of the commanders of Shah Safi's army) was ordered to move his residential site from the fortresses and to destroy them. This meant centralising the government of Kurdistan in a position that was accessible to the Iranian state, creating a gateway for the government to control the region through the local representatives. In a later period, the Ottomans did something similar in the city of Sulaimani with the leaders of Baban's family. In contrast to most local historians, Vasilyeva reckons the above date as a starting point of city formation. Although this date (1639) indicates a two-year difference from that which most local chronicles argue, it can be considered a logical reason for the foundation of Sanandaj, because the purpose of the peace treaty included defining new boundaries between two Empires (Roemer, 1968).

In line with the above view, there is evidence in the local chronicles, which to some extent justifies the view of Vasilyeva. An attack took place against the region of Kurdistan and Baghdad by the Ottomans just one year after settling the Wali of Ardalan (Sulaiman-Khan) in 1637. As this intrusion defeated the Iranian side and led to the loss of Baghdad and some parts of Mesopotamia, Shah Safi had to strengthen the central position of power in Kurdistan in order to create a strong buffer against any future attacks from his rival (Kurdistani, 2002; Roemer, 1968; Sanandaji, 1996). From this point, he forced the Wali of Ardalan to change the site of power from the mound-towns to the site of Sanandaj. This defines the change for the following year just after the accession of a new Wali to the government of Kurdistan. This also clarifies that the change of capital was not a direct decision taken by the Wali of Ardalan, but it was imposed by the Shah of Iran. From this perspective, the word of Ghamha (sorrows), "the word that was applied for the time of city foundation by Gnostics" (Kurdistani, 2002: 27), can be ascribed to the act of Wali because he was under compelling circumstances which forced him to change the site of power and create a new town under military duress.

The Period of City Foundation

From the above review, it became clear that the political importance of the city far outweighed its economic function. This means that its political dimension was the main inducement for city formation and even its later development. In line with this view and with the fact that the founders of the city, mainly Wali Sulaiman-Khan Ardalan, were aware of four centuries experience of castellated cities as a strong point for security, the core of the city was again settled on the top a hill, where it gave a convincing sense of being and provided the possibility of overlooking the surrounding area. However, the new experience of place needed to be different from the first one (mound-towns), because it was meant to project a structure for a new city. From this perspective, the place of power, in contrast with previous generations, must be settled in a position which could facilitate the structure of a medium-sized city, the city which had to be known as the capital of Ardalan in the future by merging the mobile capitals. This means that the seat of power was still the main core, but in joint structure with other parts to create a unity, namely the city of Senna-dej (Sanandaj). There are, however, many hills with different dimensions on the site of the city, for instance the large hill of Toos-nauzar, which also dominated over faraway scenery (the near and distant boundaries to the city's location) in the east, but the seat of power was located on a hill in the centre of the site, around which it was possible to structure the other elements of the city. Other factors were also important in selecting the seat of power. The main one is the sacred place of Peer-Omar close to the top of the hill which lends ritualistic importance to the place. Another factor is the existence of the remnant parts of an old Qala on the hill and a village on the hillside which gave a point of reference to the founders for settling there (Fig. 2.1).

After settling and building the main part of the city, the next important matter was the population of the city. In this regard, the review of local chronicles did not show any proper view or explanations. For example, Mardukh (1972: 48) says that "the city of Sanandaj was built based on military motivation and as a political centre to house a small number of Khans and their relatives. As the capital of Kurdistan, it attracted many professions leading to its development in the course of time". First of all, presumably, the inhabitants of other Qalas, particularly the Hasanabad Qala, were moved to settle there and construct the residential quarters and other public buildings (Vasilyeva, 2000). As the main capital of the Ardalans' power in the region, it attracted people from neighbouring villages and region to move in as well. It also drew Christians, Jews from surrounding areas.

As the issue of population was diversified, it is better to talk about the city's minorities and how they settled in the city. Again, none of the local historians did any explanation. Based on the evidence of Kinneir (1813: 144), the city contained "about eight thousand souls which of those numbers two thousand were Jews, Armenians, and Nestorians" in that period. This is a noticeable number of minorities for a middle-sized Kurdish city in that period. The roots of these people can be explained with reference to the policy of Shah Abbas (1587–1629) in "transfer of three thousand Armenians families from the Azarbaijani city of Julfa to Isfahan" (Roemer, 1968: 271). For Ahmadi,[7] the roots of these people can be traced back to two sources: firstly, the existing people on the site of Senna, and secondly, the people who were living in some of the villages in the territory of Ardalan. After the establishment of the city, especially in light of its security within the fortified walls and the control of a strong central government of Safavid, the minority populations immigrated into

[7] Interview 3 July 2003.

the city. Generally speaking, from the above perspectives and the view of Nikitine (who was the Russian Consul in 1915–1918):

> If the overambition of a Kurd is to grasp the leadership, the overambition of an Armenian is to take the position of a leader's treasurer, thus, they complete each other. (Nikitine, 1956: 178)

It is likely that the leaders of Ardalan, echoing what Shah Abbas did in Isfahan, tried to gather those minorities in order to supply the city's needs in terms of trade and commercial activities. Kinneir (1813), who entered the city in 1813, observed that those people were engaged in trade with Mosul, Baghdad and Isfahan.

Apart from the population, there are also no clear explanations in the historical records concerning the city's foundation. What they emphasised more is that Sulaiman-Khan Ardalan changed Hasanabad Qala to the site of Senna (Sanandaj) by order of Shah Safi (Mardukh, 2000; Sanandaji, 1996). In this movement, as Wali experienced the splendid city of Isfahan, he tried to build a new city similar to what Isfahan contained but with smaller dimensions. For this reason, he invited professionals such as architects, stonecutters and carpenters from Isfahan "to revive what was left from the city which had actually disappeared" and to build new mosques, hammams, madāris (schools), bazaars and bridges around the Qala (Vasilyeva, 2000: 4) and a range of qanats (aquaducts) for the main mansions of the city. To investigate how the main neighbourhoods of the city were formed, the locations of Jamia mosque (Friday Mosque), city walls and other main elements, the existing socio-spatial features of the city and some old maps are useful sources.

In the case of these neighbourhoods as the main structuring elements of the city, the etymology of the quarters' names to some extent indicates the social structure of its inhabitants. In this regard, the interests of the Khans received priority in settling on the site so that they occupied the middle part and allocated the other parts to people according to their social backgrounds. This is due to the structure of the semi-feudalist community, where the leader (Khan) is a source of both continuity and change. He is a decision-maker and a determiner of social organisation.

From this point onwards, and in reference to the old map of the city in 1851 (Mehriar et al., 1998), it can be observed that the social structure of the city was conceived according to these four areas, conveying the function of the city in that period. A physical structure comprised the core of the city as the seat of power and nobles with the dominant feature of the Qala, called Miyan-Qala, which means "the citadel in-between". The servants' quarter, adjacent to that of the nobles with easier connection to the north, was named Jawr-awa, meaning "the place developed by toilers". The neigbhourhood of the merchants, with the dominant feature of its bazaar situated towards the east of the hillside of the castle, which completely dominates over the bazaar, is named Aqa-Zaman—"the quarter of the Bazaar". The quarter of the transport organisers is situated on the other side of the river in the south and is called Qatar-chyan, "the people that used mules to transfer the goods" (Fig. 2.3). This

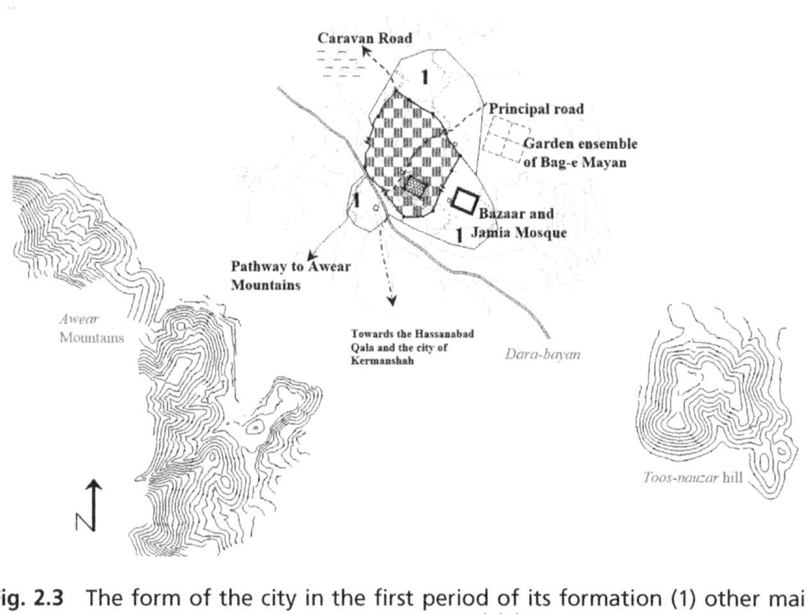

Fig. 2.3 The form of the city in the first period of its formation (1) other main parts of the city (*Source* Author, Autumn 2005. The fortified quarter of the Khans/ruling class with the Qala on the top of the hill in the south corner of the quarter)

structure is confirmed by the words of Kostof (1992: 73) who says that:

> The hilltop is where a town usually begins, with a princely citadel, a cathedral [or Mosque], a basic nucleus of feudal authority. Down in the valley is a civil core of craftspeople and traders.

The only exception is the location of the Jamia Mosque (Friday mosque) as the main element of an Islamic city. In the historical records, there is no evidence about the location of this mosque or even of its name until 1724, when the city was captured during an Ottoman invasion. Considering the two parameters of location and period of construction, there are three possibilities for the first Jamia Mosque of the city. First, the grave of Peer-Omar adjacent to the seat of power (the Qala), as it contained an additional building named Mosque of Peer-Omar. The second one is the Qula Mosque in the quarter of Qata-rchyan. The third one is Mola Ahmad Qeamati Mosque close to the principal bazaar. Although Qula mosque was recognised[8] as the oldest mosque in Sanandaj, by referring to the structure of Islamic cities in general, it is logical to consider the mosque of Mola Ahmad Qeamati as the first Jamia mosque because of its location and direct relationship with the bazaar.

Another important element of the city structure is the city walls. In this regard, apart from Vasilyeva (2000), local historical records do not mention any sign of the city walls in the first period of its foundation and even later on. Vasilyeva (2000) indicates that Sulaiman-Khan Ardalan started an impressive set of construction work, including the city walls. For the first time, signs of the city's walls were recorded by Russian surveyors in 1850 (Mehriar & Shamil, 1998) and Claudius James Rich (1836). According to the Russian document, the city included two walls, the first one enclosed the palaces/Qala and the second one embraced the inner-city or the quarter of the Khans (Miyan-Qala). Apart from this document, old photos of the city taken in 1886 show some

[8] As Bahrami (an expert in Kurdistan Cultural Heritage) and Marufi (a retired official) explain, there are only two mosques, the construction of which can be traced back to the period of the Safavid Empire: one in the quarter of Qatar-chyan named Qula mosque and the other close to the bazaar known as Mola Ahmad Qeamati mosque. Both recognised the first one as older than the second but identified the mosque of Mola Ahmad Geamati as the first Jamia mosque of Sanandaj (Interview 27 August 2003).

parts of the city walls. The imperative of this element for a city like Sanandaj is obvious, due to its location close to the frontier line of two rivalling empires: Safavid and Ottoman. Such location certainly encouraged the Wali to extend the defence elements in his capital.[9] Apart from this, the city was built as "a politico-administrative centre of Ardalan principality" and its founders had experience of the structure of mound-towns with fortified walls (Vasilyeva, 2000: 5).

We can conclude that the main structure of the city, comprising four main quarters with the two prominent features of the bazaar complex and the Qala, was formed in the first decade of its foundation. It had been continuously extended and renovated by other Walis succeeding Sulaiman-Khan Ardalan until the end of Bani Ardalan as the ruler of Kurdistan in 1867. That is why Vasilyeva (2000: 4) points out that

> In a short period of time an impressive set of building works was started and successfully accomplished: a castellated palace, mosques, madrases, bazaars, bridges, baths, city walls, garden ensemble of Bagi-I Meidan and a range of subterranean canals (Qanat) from the outside sources.

The importance of these buildings was of such an extent that "Jean Baptist Tavernier, presumably the first European who visited Sennah in the time of Sulaiman-Khan, described his enthusiastic impressions" in the following words: "La mansion ou il demeure, est une des plus belles de perse" (Vasilyeva, 2000: 4). [The house where he (Wali of Ardalan) lives is the most beautiful one in Persia].

From the above review, the city of Sanandaj should not be considered as an organic city, but rather as a semi-planned city in which the location and structure of its main parts were first conceived by the founders. The detail of each part, especially the way of building and aggregation of spatial elements across the quarters, was then left to the conventions among neighbours, due to the self-government structure of each neighbourhood.

[9] In addition to the foundation of Sanandaj with two defence walls, the city of Kermanshah, situated 80 km south of Sanandaj, was also fortified by Safavid sovereigns against the menace of the Ottoman Empire (Lockhart, 1939).

After the City Foundation

After the first period, especially after Wali Sulaiman-Khan, the Safavid government again tried to minimise the power of Ardalan's family in the region. After the death of Sulaiman-Khan, the territory of Ardalan was divided into six parts and only three parts remained for Ardalan's Emirate (Vasilyeva, 2000). Apart from the political dimension of this decision which affected Ardalan's power in the region, there was a pause in the process of city development. Due to the lack of information in the historical records concerning this change, it is difficult to provide a proper explanation about the effects of this political decision on the socio-spatial features of the city. What can be conceived is that the contents of historical records did not show any trace of major building works from the time of this action in 1656 until 1701. This means that the city experienced a noticeable stagnation in building activities for nearly half a century.

After this period, following the Safavid policy of further Iranicising the Kurdish people, the government of Kurdistan was handed over to others (four Walis) for nine years. In order to justify such political action, the Safavid Shah (Shah Sultan Hosein: 1694–1722) ordered the new Wali to do some construction work in favour of the people. For this reason, when the new Wali came to power in 1701, he built some mosques and a caravansary. As Sharafkandi (2002) indicates, a caravansary was built close to the old bazaar of Sanandaj in two storeys containing sixty chambers for commercial activities. More importantly, the caravansary was endowed to the shrines of Karbala and Najaf in Iraq in line with Shia ideology instead of the mosques and shrines of the city. The continuation of such policy in opposition to the ideology of the Sunnite people of Kurdistan[10] led to a popular uprising and eventually the dismissal of the Wali. This action convinced the Safavid court that the foreign Walis could no longer govern the Kurdistan region. For this reason, they again invited the Ardalan family to continue their authority in Kurdistan.

---

[10] As Kurdistani (2002: 101) and Mardukh (2000: 339) indicate, when the second Wali, out of Ardalan's family, came to power and he realised that the shrine close to the site of power was named the shrine of Peer-Omar, he destructed it because its name recalled the name of the Sunnite second caliph.

Although the government again was transferred to the Ardalan family, the historical records do not mentioned any important building work until 1720, when the city was conquered by the invading Ottomans (Babans' family, the rival of Ardalan's family) as the Safavid Empire collapsed. As local historians indicate (Kurdistani, 2002; Mardukh, 2000), the Walis of Ardalan were mostly engaged in wars against internal and external invaders, especially the Afghan invasion in eastern Iran. Therefore, they did not devote any energy to building activities and the city was no longer the secure place for its hinterland. This certainly influenced the city development in that period.

The city of Sanandaj was conquered by Khan-e-Pashah Baban in 1720. This gave an opportunity to the Baban family to put their imprint on the city's spatial features. This took the form of a new mosque with two tall minarets and its complementary element of Madraseh (school), using money supplied by the Ottoman Sultan. Although no historical records referred to this mosque as the Jamia one, from an Ottoman policy perspective (in opposition to Safavid ideology and as a representative of the Sunnite sect) as Mardukh (1972, quoted in Sharafkandi, 2002: 124) claims, "in the minaret of this mosque the Izan (call to prayer) was performed", it can be realised that the mosque was a new replacement for the city's Jamia mosque. More importantly, it was erected on the slope of the Qala hill just adjacent to the fortified wall of the Qala facing the bazaar, the site of Asnaf (guilds) and merchants.

Apart from the political challenge of the Ottoman Empire against the Safavids, the erection of this mosque in front of the Qala's view towards the east (the Qala having been the main seat of Ardalan's power, and the east as the predominant direction of the city) can be conceived as a monument of the Babans' prevalence over the Ardalan family due to their long military and political competitions. The reign of Baban lasted ten years until Nader-Shah (1736–1747) appointed the government to Subhan Werdi-Khan (another leader of Ardalan) in 1730. As Mardukh (1972, quoted in Sharafkandi, 2002) indicates, the people of Sanandaj were more interested in Iran, which is why they asked Nader-Shah to recapture the city.

As Vasilyeva (2000) indicates, there were two different attitudes towards their capitals between these leaders (Ardalan and Baban). As the city of Senna was the symbol of Ardalan's power, they spent much effort to renovate and decorate the city in the course of time. On the contrary, the Baban leaders spent their efforts on portable things because "they were less interested to underline their social primacy for their power rested on the tribal community" (Vasilyeva, 2000: 6). This can be witnessed in the words of Rich (1836: 214, 215), who entered the city of Sanandaj in 1820:

> One of the Khans observed, when I praised some of the buildings, that they spent all their money about their houses, but that the Bebbehs (Babans) invested their property in money and transportable effects, that they might be ready to fly at a moment's warning, in case of trouble or the deposition of their chief.

For this reason "the city of Sulaimania, the capital of Babans, was more decent than Sennah", it was the city of Sennah which owned the royal splendour—"une allure royale"—as "its constructions contributed to the splendour of prince's power" (Nikitine, 1956 quoted in Vasilyeva, 2000: 6).

When the Zand dynasty (1747–1787) arose, the city of Sanandaj experienced great problems of war and devastation. In the first years of the Zand dynasty's establishment, a clash took place between the heads of this dynasty and the Wali of Ardalan (Hassan Ali-Khan) (Perry, 1968: 67). As the army of Zand was defeated by the Ardalans, that created a sense of revenge against the Ardalans' family in the mind of Karim-Khan Zand. For this reason, when he approached the city in 1750 "his army sacked and burned" most of the city. As Sharafkandi (2002) indicates, on this occasion, Karim-Khan Zand settled in Sanandaj's Jamia Mosque. From there, he ordered his troops to burn the city.

After this event, which led to the destruction of the city and the displacement of the people, another Wali of Ardalan took the power and tried to reconstruct the city in 1757. Similar to the other periods, there are no clear explanations in the historical records about the circumstances of the city's reconstruction. Based on the local chronicles, when Khusraw-Khan II was appointed as the Wali of Kurdistan, the people turned back to the city and Wali started an impressive programme of building works. He first changed his seat to the Hasanabad Qala in order to reconstruct the Sanandaj Qala. In reconstructing the Qala, he built a new hall (Balakhanih) in an eastward direction named Khusrawieh (Mardukh, 2000: 358). He also built a new Char-baq (four gardens) named Baq-e Mayan in the north-east of the city by "using the specialist engineers". This area later became the Quarter of Char-baq. As discussed, the notion of this Baq had been conceived and to some extent projected by Khan Ahmad-Khan before the birth of Sanandaj in 1616–1637 (Fig. 2.3). As the idea and project of this garden emanated from the Char-baq plan of other Iranian cities, such as Shiraz and Isfahan, and it was created by engineers, it was well created. Masturah-Khanum (quoted in Sharafkandi, 2002: 168) describes it in the following words:

> When I was child, I usually went to this garden for the sake of recreation. It was really a well-defined garden not other thing. The streams across the trees recalled us the sense of heaven which made it unique in the world.

The reign of Khusrau-Khan coincided with the rise of the Qajar dynasty (1794–1925). Following the brief rule of two successors, the reign passed to Amanullah-Khan the Great, the son of Kusrau-Khan, in 1799. His long tenure of office is a remarkable period in Ardalan's family, and a period of royal splendour for the city. This period was described by historians as an important period, featuring close collaboration with central government which eventually led to the demise of the state-like status of the Ardalan emirate (Hambly, 1968; McDowall, 2001). Considered from the vantage point of city development, it however must be recognised as a period of impressive building works which represent a strong image of the city as a place of royal splendour in the perception of foreign travellers. As Mardukh (2000) explains, Amanullah-Khan spent more efforts in the development of Ardalan's emirate. He built new elements in the site of power/Qala (the interior mansion of the Qala, the court of Gulastan, the hall of Delgusha and a square in front of the Qala), in the surrounding areas (the site of Kani-Shafa on the slope of the Awear Mountain) and even in some rural areas. As this author states, the most important mansions of the city are traced back to this period including

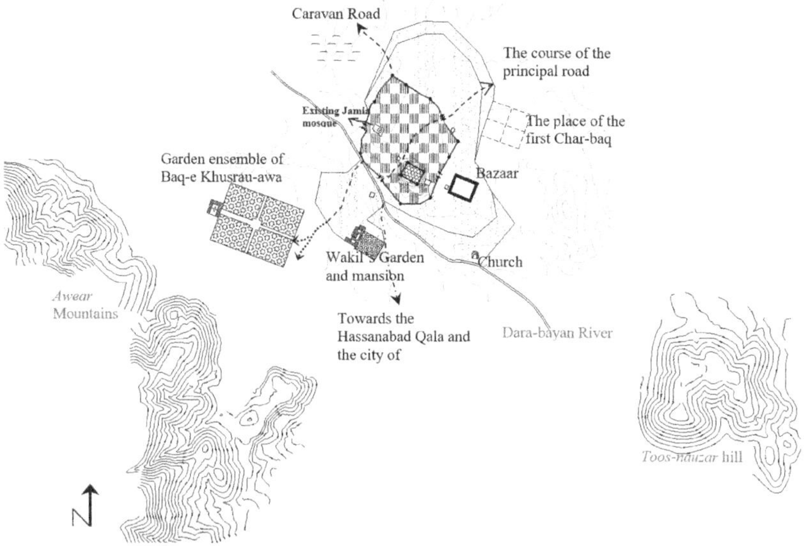

Fig. 2.4 The second phase of city development in the first period of the nineteenth century: formation of the Second Ensemble of gardens (Khusrau-awa Baq) and the existing Jamia mosque (*Source* Author, Autumn 2005. The Qala/citadel and the place of the second Jamia mosque [Baban's mosque] of the city in its east side overlooking the bazaar. The fortified quarter of Khans/ruling class with two main elements of the existing Jamia mosque and the Qala/citadel. The bazaar complex accompanying the first Jamia mosque of the city; Mola Ahmad-e Geamati)

the current building of Jamia mosque and a new Char-baq north-west of the city, opposite side of the first Char-baq (Fig. 2.4).

Generally speaking, the portrait of the city in that period can be imagined based on the following words of Kinneir (1813: 199) and Rich (1836: 144):

> Secluded in the bosom of a deep valley, well-cultivated and interspersed with orchards of peach, apricot, pear, apple, and cherry-trees, Senna is, at once, a most romantic and flourishing little town.

> Sinna wore a much more imposing appearance than I expected, with its castellated place on the height, and some good-looking buildings round the foot of it.

Such impressive image of the city can be ascribed to the power of Wali and to the viewpoint that "the fortunes of a particular town or district at any given time depended to a very large extent upon the character and competence of the local officials" (Hambly, 1991: 562).

From the above review, the two spatial elements of the new Jamia mosque and new Char-baq gained special character and identity in the city's history. Based on the local chronicles, there are two important reasons behind building a new mosque in that period. One relates to the political impulse against the monument of the Ottoman Empire (the mosque of the Babans). As Mardukh (2000: 344) and other local historians explain, the existence of the physical imprint of the Babans was unbearable for the Wali, as it was erected in front of the Qala's view towards the east. For this reason, he first destroyed and replaced it with a garden named "Baq-e Ferdusi". Then, he erected a new mosque close to the western gate of the city. The second reason is still directed against the policy of Ottomans, but with religious purposes. The Wali of Ardalan aimed to build a new large mosque comprising both functions of praying and education as a main centre of Ulama[11] in Kurdistan.

Concerning the second reason, Ahmadi[12] and Bahrami[13] explain that "the purpose of Wali was to create a centre for Ulama (the religious authorities) in the region to comprise the whole processes of learning of that period". As they explain, before this centre was built, most students of the Sunnite Shafiai sect had to complete their studies (especially the final stages) in the Iraqi region under the control of the Ottoman Empire. From this point, building a new centre of teaching overcame that gap in the education of the city and the whole region. This centre was of such great importance that it attracted many Ulama and students from other parts of Iran and the Iraqi region. For this reason, in the period of

[11] In Islam, the ulama are the guardians, transmitters and interpreters of religious knowledge, including Islamic doctrine and law.

[12] Interview 29 July 2003.

[13] Interview 28 July 2003.

Amanullah-Khan, the city of Sanandaj was known as "Darull-Elm" (the house of learning) or "Darull-Adab" (the house of training), which made it unique in the west of Iran (Kurdistani, 2002).

Although the act of Wali, in changing the place of the Jamia mosque, led to the development of a well-known city in the region, it further secluded the public place of the main mosque from the foci (the bazaar complex) and from the commoners by maintaining it in the quarter of the ruling class (Fig. 2.4). This means separation of the Jamia mosque from the public and placing it in the realm of nobles. Keeping this fact in mind and considering the previous status and locations of the Jamia mosque, it can be claimed that the life of the Jamia mosque defines the life of the city.

> The mosque [Friday one] is the city, or rather the focus of the city, and its physical ambiguity represents a definite spiritual fact and purpose. It records the city's life and growth and comprises many seculars and all its spiritual functions. (Pope, 1967 quoted in Kheirabadi, 1991: 65)

Concerning the new element of Char-baq, the historical records have not explained any particular socio-political reasons behind its creation. For example, in the words of Mardukh (2000), it was built in honour of the Wali's son, Khusrau-Khan the second. And Rich (1836: 200), who visited the garden in the period of the Wali, explains that

> The garden was planted by Amanullh-Khan, the present Wali, about fourteen years ago, and was named by him Khusrooabad, in honour of his father, Kuosrau-Khan, the late vali.

Presumably, the words of Rich seem to be the right explanation for the creation of this garden. As explained, Khusrau-Khan, the father of Amanullah-Khan, created the first Char-baq in the north-east of the city. Based on the view of Masturah-Khanum, who lived in the period of Amanullah-Khan as the wife of his son, this Bagh had lost its status as a royal garden (cited in Sharafkandi, 2002). Consequently, Wali tried to revive the idea of his father in the form of a new Bagh in a new location for two purposes: first, reviving the memories of his father concerning the first Char-baq, and second, putting his imprint on the city's spatial structure to crystallise his power for future generations.

The last reason can be highlighted regarding the challenge of Wali with Wakils' family. As historical records indicate (Vasilyeva, 2000: 7; Mardukh, 2000), "the Wakils' family had been continuously opposite to the Wali and acted on the side of Ardalans' adversaries". For this reason, they were driven out of the ruling-class quarter and settled in the quarter of Qatar-chyan. As they created a site of power for their families in that neighbourhood (Fig. 2.4), it can be imagined that the creation of Khusrooabad Bagh was also a physical response against this family and their sense of power in the territory of Ardalan. From this point, we can rely on the view of Grabar (1984: 65) who recognised architecture as an "expression of power". Apart from the political dimension, it can be justified by its function as a summer house of Wali on the slopes of the Awear Mountains. This is an explicit purpose, which is also clear in the words of Asaf[14]:

> During the summer, as the weather of Sanandaj, especially the houses close to the river of Dara-bayan, got warm with growing of mosquitoes, it was really difficult to endure its conditions. For this reason, the Wali and his relatives usually changed their site to Khusrau-awa Baq close to the slope of Awear which possessed fresh air with a chilly breeze.

From the above review, the general structure of the city in that period, with the exception of the new Jamia mosque and Char-baq, is to some extent what the city projected in the first period of its foundation. As Rich (1836: 207) indicates, the city of Sanandaj presented the following spatial structure:

1. "the castellated palace, situated on a high mountain, which overlooked all the town by "maintaining a degree of state and splendour, superior to anything I have seen in Persia, except at court" (Kinneir, 1813: 144, 145), and was surrounded by a wall;
2. at the foot was another part of the town and another fortified wall;

[14] Interview 11 July 2003.

3. beyond this was more of the city which again was surrounded by a wall".

Although Rich's description exactly clarifies the city's main parts, his last point must be modified, since the city only contained two fortified walls, not three: one around the castellated palace, and a second around the ruling houses. As explained before, the map of Russian surveyors drawn in 1850 exactly shows the lines of two fortified walls around two parts of the city while the third part is free of any man-made enclosure.

Apart from the physical descriptions of the city, the accounts of European travellers, which mostly took place in the period of Amanullah-Khan, have some estimates of the population. Kinneir (1813), who approached the city in 1813, estimated its inhabitants at about eight thousand souls. Rich (1836), who visited the city in 1820, on the other hand estimated the population of Sanandaj to comprise four to five thousand families, "that is equal to twenty or thirty thousand people". This shows a great difference in only seven years, the period of Wali Amanullah-Khan, that is, the period in which the city reached the zenith of its development. As Masturah-Khanum (quoted in Sharafkandi, 2002: 233) indicates, "the prevalence of pestilence killed more than eight or nine thousand people" in 1830. This means that the number of people who died is more than in Kinneir's estimation. For this reason, Rich's estimate of the population is more reliable. The other evidence which again is in line with the view of Rich is the estimate of Russian surveyors in 1850. They numbered the population of the city at about 10,240 people.

Apart from the whole population of the city, they also estimated the population of the religious minorities in the city. For Kinneir, of eight thousand souls of the city two thousand are minorities, and for Rich, of four to five thousand families, 200 families are Jews and fifty houses belong to Catholics. Although the number of minorities are nearly the same in both estimates, they show a great contradiction concerning their proportion to the whole population. These estimates show that the minorities formed a high percentage of the city's population in that period. They settled in the Bazaar neighbourhood, which embraced the bazaar complex. They mostly traded with other cities in both Iran and

Iraq: Mosul, Baghdad, Diarbakr, Isfahan and Tehran (Kinneir, 1813; Rich, 1836). Being both merchants and minorities affected the spatial structure of the quarter so that it had a particular spatial pattern distinguishable from other quarters. It served to strengthen their security and consisted of a mazed pattern of street networks with small culs-de-sac to disorient invaders (Fig. 3.8).

The Fall of the Splendid City

The following period of Wali Amanullah-Khan (1799–1824) gradually witnessed the fall of the Ardalans' governorship in the region. It also led to the demise of the splendid image of the city. This means that "as the symbol of Ardalans' power Sanandaj followed all the curves of their political fate" including continuous efforts of this family to rebuild and decorate its constructions (Vasilyeva, 2000: 5).

As explained, the state-like status of the Ardalan emirate was diminished when the rulers of the Qajar dynasty followed the policy of intermarriage launched by Shah Abbas the Great. This happened when Fath Ali Shah's daughter (1797–1834) got married to Wali's son, Khusrau-Khan the Third, in 1819. This event led to a change in the custom of hereditary assignment of power, to such an extent that the younger son, Khusrau-Khan the Third, was appointed to be the successor of Wali Amanullah-Khan instead of the elder son. From this time onwards, as Lambton (1968: 467) points out, "the province of Ardalan was brought under a greater degree of control by the central government". It means that Ardalan's emirate lost its state-like status. This trend gradually resulted in a weak role for the Ardalan family in the region, and even in a decline of the city's development. This, alongside the devastating prevalence of pestilence, which in 1831 caused the death of more than nine thousand people, created a great problem for them to manage the city's development properly. Its impact was such that the review of historical records did not show any trace of building work in that period until 1846, when the last Wali of Ardalan (Amanullah-Khan the second) came to power by the order of Naser o-Din Shah (1848–1896).

The period of the last Wali included the particular socio-political event of the building of a new mosque, named the Wali's mosque, close to the site of the bazaar on the slope of the Qala's hill in a low-lying part of the site of Baban's mosque. As Mardukh (2000) explains, this mosque was created by Wali Amanullah-Khan the Second in favour of Merza Karim-Khan-e Hammadani, the founder of the Sheikite religious school which was in line with Shiite ideology. Due to the social cataclysm against this act of Wali, which was mostly organised by Bazaaries and Ulama, he changed his mind and left the construction of the mosque unfinished. The Wali attacked one of the bazaar's caravansaries in order to pacify the disturbance. This event was recorded as "the War of Caravansary" (Mardukh, 2000: 405).

Based on the survey of the city in that period (1850) by Russian surveyors, the structure of the city is to some extent clear and can be summarised as follows:

1. The fortified Qala located in the centre as the oldest part of the city.
2. The quarter of the ruling class with fortified walls and four gates towards the cardinal points of the campus located behind the Qala in the western part of the city.
3. The bazaar complex located at the core of the city, surrounded by the compact form of the houses in the Bazaar neighbourhood.
4. The Jamia mosque set back from the main foci of the city as it was maintained in the quarter of the ruling class, close to the western gate of the city.
5. The other neighbourhoods encircling the city's main elements without any wall as the third part of the city.

In this period, the only change that took place in the spatial structure of the city is in relation to the new neighbourhood named Char-baq, created on the site of the garden ensemble of Bag-e Maidan, the first Char-baq of the city located in the north-east part of the old city. In the period of the last Wali of the Ardalans, therefore, the city comprised five main quarters: the central quarter of the nobles, surrounded by a second fortified wall and the other four quarters as the places of commoners without any man-made enclosure (Fig. 2.4).

The reign of Wali Amanullah-Khan the second lasted until 1867. As Sharafkandi (2002: 278) says, when the last Wali of Ardalan's family died in 1867, there was no strong agreement among them for transferring the power internally. The Shah of Qajar, Naser o-Din Shah, had been looking for an opportunity to break the power of Ardalans' family. He used it and appointed his uncle as the ruler of Kurdistan. This was recognised as the end of Ardalan's government over the region of Kurdistan in general (Ghandchi, 1999), and the city of Sanandaj in particular. From this time onwards, the government of Kurdistan never again passed to local rulers.

The above period can be recognised as a remarkable time in the history of the city, after which the city entered a period of decline in its socio-spatial development. As can be seen from historical records, the non-local governors mostly engaged with military actions in order to pacify the disturbance of some tribes in the frontiers, or the protests of people against them. For this reason, they mostly spent their efforts and money on fortifications to stabilise their power[15] in the area and to appease the people's wrath (Mardukh, 2000).

Following this period, the territory of Kurdistan was annexed to the province of Kermanshah by the decision of Naser o-Din Shah in 1874. This change caused the city of Senna to lose its status as the capital of Kurdistan. It also affected the city's development, because the main concerns of new governors were usually in line with the capital city and not with secondary cities. In contrast to the previous periods, the governor of Kurdistan was appointed by the governor of Kermanshah and not by the central government. Due to their non-local roots, the new governors had no interests in and motives for city development and most of them seemed to have been venal, weak, or a combination of both. They were usually changed after one or two years due to their weak management and the continuous protests of the people against them.

After the reign of Ardalan's family, the city witnessed thirty-five governors during a period of only 64 years, from 1867 to 1931.[16] This means

---

[15] A new governor created a site for military purposes (barrack) in front of the Qala in that period (Mardukh, 2000: 423–427).

[16] On the contrary, the reign of Ardalans lasted for 230 years with 38 Walis. It means that each Wali governed the region for nearly six years.

that each governor maintained his power for less than two years. In such a short time, as explained, they could only manage their affairs concerning any rebels in their territory. Therefore, the change of power from local to non-local governors caused negative impacts on the city's prosperity. As Mardukh (2000) and Kurdistani (2002) indicate, even the splendid Qala was gradually destroyed in the course of time and it only lasted until 1919. Most of the valuable columns of bright marble and its famous stone pool were transferred to other cities by the non-local governors. More importantly, the local historians did not mention any particular building activities in this period. This recalls the notion of Hambly (1991: 562) that the well-being of traditional Iranian cities in the period of the Qajar dynasty was inextricably linked with the actions of successive governors, especially the governors who had a link with the local people. That is why, in the view of Vasilyeva (2000: 9):

> Sanandaj had a unique atmosphere which was created and maintained by the Kurdish habits and traditions many of those were abolished by the governors after the removal of the Ardalans.

Conclusion

Through analysis of the city's historical background, this section has explained the city's foundation and its development in the course of time. It was found that the main reason for its building lies in the essence of political challenge between two empires in the seventeenth century, one with the motive of Sunnite ideology located to the west of the Zagros Mountains (Ottoman Empire), and the other with the purpose of extending the Shia religion to the east of these mountains (Safavid Empire). In this challenge, the luck was with the eastern side as it emphasised the ancient associations of Kurds with Persia. From this point, the city developed according to Safavid policy, as a politico-administrative centre and strong bulwark against the Ottoman rival by moving down the Kurdish people from mound-towns to relatively level areas in the current location of Sanandaj. In so doing, the Safavid Empire took a great step in Iranicising the Kurdish people and breaking down their semi-independent status on the Perso-Turkish frontiers.

It was also found that the city's destiny was completely bound with the power of Ardalan's family. The city of Sanandaj thus became a symbol of this family in following all the changes in their political fate. Although the local historians did not mention any trace of building works in some periods, it is clear that the city was rebuilt and renewed many times by the efforts of the Ardalan family.

References

Alizadeh, H., Bahrami, B., & Farahani, L. M. (2019). Geophysical setting and urban form of a traditional city: Examination of the Kurdish city of Sanandaj, Iran. *Proceedings of the Institution of Civil Engineers-Urban Design and Planning*, *172*(6), 237–247.

Ardalan, S. (2008). *Kurdish Tribe of Ardalan: Iran and Ottoman Empire conflicts* (M. Ardalan, Trans.). Iran History Publication.

Babani, A. (1998). *Saerull-Akrad dar Tarikh va Guqkraphia-E- Kurdistan* (An overview of history and geography of Kurdistan). Awrang Press.

Bayat, A. (1988). *Kuliate Guqkrafia- E- Tabeeai va Tarikhi Iran* (A historical and physical geography of Iran). Sepehr Press.

Boyce, M. (2001). *Zoroastrians: Their religious beliefs and practices*. Routledge.

Dultshahi, A. (1984). *Guqkrapgiai Qkarbi Iran ya Kuhhai Nashnahkti Evista*. Naghshi Jahan.

Ghandchi, S. (1999). *Kurds and formation of central government in Iran*. http:// www.ghandchi.com/02-Kurds/KurdsEnglish.htm. 2003.

Grabar, O. (1984). Palaces, citadels and fortifications. In G. Michell (Ed.), *Architecture of the Islamic world: Its history and social meaning* (pp. 65–79). Thames & Hudson.

Hambly, G. R. G. (1968). Agha Muhammad Khan and the establishment of the Qajar dynasty. In W. B. Fisher (Ed.), *The Cambridge history of Iran* (Vol. 7, pp. 104–174). Cambridge University Press.

Hambly, G. R. G. (1991). The traditional Iranian city in the Qajar period. In P. Avery, G. Hambly, & C. Melville (Eds.), *The Cambridge history of Iran: From Nadir Shah to the Islamic republic* (Vol. 7, pp. 542–589). Cambridge University Press .

Habibi, S. M. (1996). *Az Shar Ta Shahr* (Analytical review of the city concept and its physical image in the course of time). University of Tehran.
Izady, M. R. (1992). *A concise handbook: The Kurds*. Crane Russak (Taylor & Francis).
Kamandi, A. (2014). *Pas Koochahaee Sanandaj* (Sanandaj's alleys). College Scientific Publication (in Persian).
Kheirabadi, M. (1991). *Iranian cities: Formation and development*. University of Texas Press.
Kinneir, S. J. M. (1813). [Republished 1973] A Geographical memoir of the Pesian Empire: Journey through Asia Minor, Armenia, and Koordistan in the years 1813 and 1814. Arno Press.
Kostof, S. (1991). *The city shaped: Urban patterns and meanings through history*. Thames & Hudson.
Kostof, S. (1992). *The city assembled: The elements of urban form through history*. Thames & Hudson.
Kurdistani, A. V. (2002). *Hadigha-E-Nasriah dar Guqkrafia VA Tarikh-e Kurdistan* (History and geography of Kurdistan) (Persian). Tavakouli.
Lambton, A. S. (1968). Land tenure and revenue administration in the nineteenth century. W. B. Fisher (Eds.), *The Cambridge history of Iran* (Vol. 7, 459–505). Cambridge University Press.
Lambton, A. S. (2011). Administration and social organisation. In E. Yarshater (Ed.), *Encyclopaedia Iranica: Cities*. Columbia University and Bibliotheca Persia.
Le Strange, G. (1905). *The lands of the Eastern caliphate: Mesopotamia, Persia, and central Asia, from the Moslem conquest to the time of Timur*. Cambridge University Press.
Levine, L. D. (1974). Geographical studies in the Neo-Assyrian Zagros - I I. *Persian Studies, 12*, 19–123.
Lockhart, L. (1939). *Famous cities of Iran*. W. Pearce & Co.
Mardukh, M. (1972). *Tarikh-e- Kurd VA Kurdistan* (The history of Kurds and Kurdistan). Qkariqhi (in Persian).
Mardukh, M. (2000). *Tarikh-e- Mardukh* (The history of Kurds and Kurdistan). Karang.
McDowall, D. (2001). *A modern history of the Kurds/translated by Abrahim unis*. Paniz Press.
Mehriar, M., Fatholayouf, S., Tehrani, F. F., & Qaderi, B. (1998). *Asnad-e-Tasveri shahrhaei Irani* (The old maps of Iranian cities). Shahid Beheshti University Press.

Minorsky, V. (1943). Sanandaj. In M. T. Houtsma, A. J. Wensinck, H. A. R. Gibb, W. Heffening & E. Levi-Provencal (Eds.), *Encyclopaedia of Islam*. E. J. Brill and Leyden & Luzac & Co.
Mizanian, N. (2017). Mastoureh Ardalan, a rarity of her time, the first woman historian. *Tehran Times*. Mehr News Agency.
Nikitine, B. (1956). *Les Kurdes. Etude sociologique et historique*. Imprimerie nationale, Bibliothèque C. Klincksieck.
Perry, J. (1968). The Zand dynasty. In W. B. Fisher (Ed.), *The Cambridge history of Iran* (Vol. 7, pp. 63–103). Cambridge University Press .
Rich, C. J. (1836). *Narrative of a residence in Koordistan, and on the site of ancient Nineveh: With journal of a voyage down the Tigris to Bagdad and an account of a visit to Shirauz and Persepolis*. J. Duncan.
Roemer, H. R. (1968). The Safavid period. In W. B. Fisher (Ed.), *The Cambridge history of Iran* (Vol. 6, pp. 189–350). Cambridge University Press .
Safrastian, A. (1948). *Kurds and Kurdistan*. The Harvill Press LTD.
Saleh, M. A. E. (1995). Al-Alkhalaf: The urban history of a traditional settlement in Southwestern Saudi Arabia. *Habitat International, 19*(1): 29–52.
Sanandaji, M. S. (1996). *Tuhfa-ye Nasiriye dar Tarikh va Jughrafiya-yi Kurdistan* (Kurdistan history and geography). Amir-Kabir.
Shapur Shahbazi, A., & Edmund Bosworth, C. (1990). Capital cities. *Encyclopædia Iranica*. Online edition. Vol. IV, Fasc. 7
Sharafkandi, A. (2002). *Mejwi Ardalan* (The history of Ardalan). Taze Nghah Press.
Soltani, A. (2006). The Sharafnama of Bitlisi manuscript copies, translations and appendixes. *The International Journal of Kurdish Studies, 20*(1–2), 209.
Vaceq Abbasi, A., & Abdi, M. R. (2017). TUS, the golden footed. *The Journal of Epic Literature* (PAZHUHESH-NAMEH FARHANG-O-ADAB), *13*(1 [23]), 89–113.
Vasilyeva, EI. (2000) The Kurdish city of Sanandaj: Ardelan's Capital. Paris, conference presentation.

3

Morphological Components

Abstract This chapter examines the city's morphological components through contextual analysis in two sections that together make clear the socio-spatial rules underlying the city's morphological components. In the first section, the city's built form within its macro and micro-geographical settings is examined to reveal the nature of the interaction between man-made and natural environments and to identify the features of the city that responded to the natural landscape. The second section examines the city's constituent elements, especially the morphological elements of social communication and spatial articulation. It provides a dynamic perspective, revealing the city's fixed points, details and the logic of urban configuration to identify those that have modelled the urban fabric. In the chapter, city walls, residential neighbourhoods (*Mahallaha*) and public spaces are discussed. Based on the results of this chapter, it is possible to say that the city's founders had perceived its main spatial structure in advance as a whole with the design principles serving the interests of the Khans/ruling class. This means that the concept of the planned city can be applied for Sanandaj.

H. Alizadeh, *Building a traditional Kurdish City*,
https://doi.org/10.1007/978-981-16-3634-9_3

Keywords Morphological components · Geophysical setting · City structure · Neighbourhoods (*Mahallaha*) · Public places · Bazaar

Introduction

After discussing the historical background of the city and making clear the factors that motivated the foundation and development of the city over the course of time, next, the built form of the city is examined in two parts. First, the geophysical setting of the city within its macro- and micro-geographical settings is examined, considering the viewpoints of Norberg-Schulz (1971) and Spirn (1988). For Norberg-Schulz, nature is a base for human orientation and his/her existence; thus, it constructs the primary components of a city. In the viewpoint of Spirn, however, the city is "a place where natural forces pulse and millions of people live - thinking, feeling, dreaming, doing, our habitat was ordered primarily by nature's processes" (Spirn, 1988: 108 and 110). That is why Kostof (1991) believes that it is impossible to isolate urban experiences from the effects caused by the earth (Kostof, 1991). To go through these leading points, analytical drawings are used to discover the nature of these interactions between the two environments (man-made and natural) and, more importantly, to identify the features of the city which responded to the character of the natural landscape.

The second part examines urban form components based mainly on the work of Kostof (1992) to understand the city's structuring elements. He begins with the concept of edge concerning the city's walls, continues with the residential quarters and then the public places. These two main sections act as a contextual morphological analysis, which makes clear the general socio-spatial rules underlying the city's morphological components. The results of the city's historical background, especially the maps and drawings illustrating the evolution of the city, are helpful to illustrate these components.

Regional Structure

> The secret of form lies in the fact that it is boundary; it is the thing itself and at the same time the cessation of the thing, the circumscribed territory in which the being and the No-longer-being of the thing are one and the same. (Simmel, 2007: 73)

The notion of "place" implies an inside and an outside character. It is meaningless to imagine the city itself without its regional context (Norberg-Schulz, 1971). The city and its surrounding landscapes are interdependent (Boardman, 1978), and natural elements are the primary components of a city. "It is in relation to nature that places and objects take on meaning". This is why nature is recognised as of key importance for human inspiration and it is in relation to nature that places and objects take on meaning (Jiven & Larkham, 2003: 71).

While investigating Sanandaj's regional context, Ardalan and Bakhtiar (1973) argue that Iran's mountain ranges (Alburz and Zagros) perform as walls defining a setting in which historic cities and towns have developed. They lend a macrostructure to the landscape by the chains of the Zagros and Alburz mountains as mighty extensive walls in the area with those of the south and east being smaller, demarcating the geographical settings of vast dimensions. This physical form of the region can be imagined as big bowl-like enclosures (Fig. 3.1) with a high outer (western and northern) rim surrounding its main regional space: "human settlements are primarily situated on the aprons of majestic mountain ranges as centripetal nodes that often relate centrifugally to a geographic space of vast dimensions" (Ardalan & Bakhtiar, 1973: 13). The mountains have also been a source of water. In the arid zone, water is usually transferred to the settlements through Qanats or subterranean canals leading from the slopes of the mountains. The signs of canals are like "pot holes left from initial excavations and the subsequent repairs" (Ardalan & Bakhtiar, 1973: 13).

Finding a secure setting seems to have had more influence than usual on the distribution of human settlements and related activities in this region. Mountains can be viewed as places that "make the structure of being manifest" meaning that "man wants to make the natural structure more precise" (Norberg-Schulz, 1980: 17, 25). To achieve this, they built what they have seen: enclosures, where nature delimits, gateways, where nature suggests entrance and paths where the setting promotes direction (Kite, 2008). That is why the bowl-shaped physiography of the Iranian plateau has had a major effect on the location and morphology of Iranian cities (Arefian & Moeini, 2016; Kheirabadi, 1991). For Kurdish people, mountains, especially the Zagros, represent the most important natural phenomenon that has shaped the Kurdish history, people, tradition and culture to such an extent that Kurdish domains end where the plains begin (Izady, 1992: 13). Therefore, human settlements within the Zagros Mountains can be viewed as isolated, self-reliant and inward-looking communities with few external contacts (Grandmaison, 1977) (Fig. 3.1). In this setting, the mountains act as points of reference for arrival and departure and as gateways for defence purposes. The historic city of Sanandaj was, in fact, a defensive gateway for the Persian Empire against the intrusion of the Ottoman Empire from the west (Alizadeh, 2004).

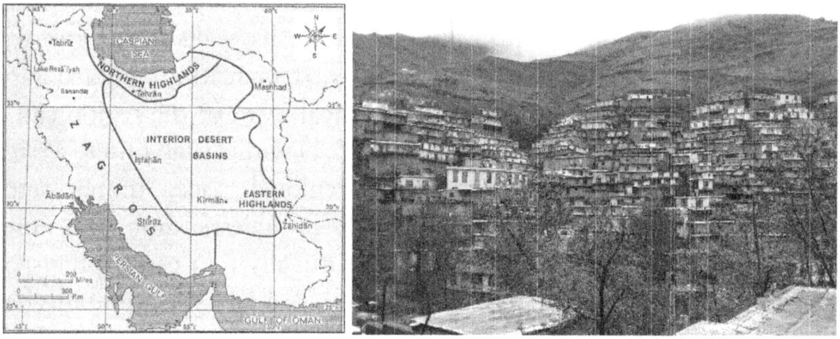

Fig. 3.1 Iran's physiographical units and a typical settlement in the Zagros Mountains (*Source* Map from Fisher [1968: 7], photopgraph by author, November 2018)

The Structure of the Natural Landscape

> Nothing exists without an edge; no temple, no person, no mere object is without its limits; everything must end somewhere for its identity to be apparent. (Leatherbarrow, 1993: 37)

On a microscale this limit is the inter-mountainous setting amid the Zagros, where the Zagros folds around the resistant blocks that form the plain of Hamadan (Fig. 3.2). The local setting comprises a relatively large bowl-like area including many hills of different sizes and different settlements (the city of Sanandaj and other rural settlements close to the course of the river). The river Cham-e Gheshlagh, flowing in a north–south direction, furthermore defines the north-west and south-west boundaries of the region. This geophysical rim can be imagined as two strong curved walls, the western part formed by the Awear Mountains and the eastern one by the heights of Salawat-awa, cut through by the Cham-e Gheshlagh River (Fig. 3.3). Apart from the city of Sanandaj, there are a number of villages alongside the course of the river—the main regional corridor and water source.

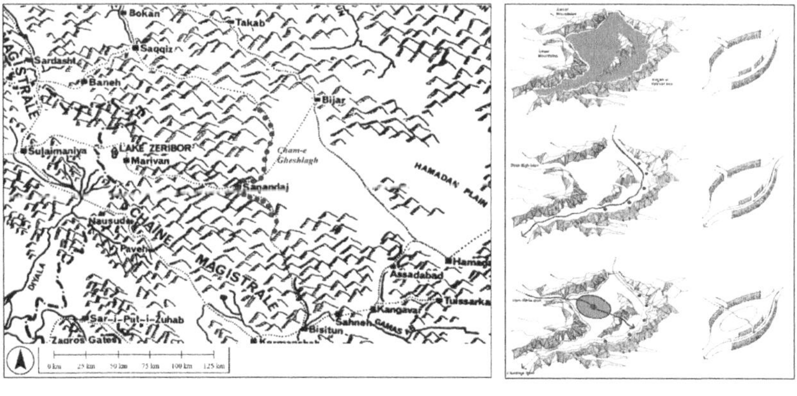

Fig. 3.2 The city's location amid the Zagros Mountains and the city's geophysical setting (*Sources* Left from Levine [1974: 101] with some adding points by author, Autumn 2005, right adapted from Ardalan and Bakhtiar [1973] and redrew by Bakhtiar Bahrami, Spring 2019)

Fig. 3.3 The sacred hill of *Toos-nauzar* as point of reference for city's structure and the Awear Mountains as the western rim of both the first and second realisation of the city's geophysical setting. They are cut through by the valley of Haft-asiaw (seven mills), thus forming two Awears, Bouchek/small and Gawrah/big. Sources: left by Mr. Kasra Ketaboullahi, Summer 2020 and right from the collection of Mr. Gazarani, date unknown

As we get closer to the city, the western rim—the Awear Mountains—is the major natural border limiting the expansion of the city to this side. The only interruption is a small valley (Doule Haft-asiaw) between Awear Bouchek (the small mountain range) and Awear Gawrah (the large mountain range) (Fig. 3.3), in which Seven Mills supplied the city and its surroundings. The eastern rim is defined by two other ranges of hills, lower in height compared to Awears Bouchek and Awears Gawrah, separated by a river (Dara-bayan) beginning from the slopes of the Awear Mountains and joining the Cham-e Gheshlagh River. These heights can be imagined as natural defensive walls that separate the city from its surrounding environments. While the city was shaped behind these heights, two paths in gorge areas overlooking the main corridor served for arriving and departing. The city is not visible to a stranger (visitor or traveller) from the main corridor. Besides these heights, there are other small surrounding hills that limit the views of Sanandaj even from the summit of the two paths from the main corridor. The corridor along the Cham-e Gheshlagh River (Fig. 3.2) makes a connection and a longitudinal axis for access to the city and neighbouring settlements.

Alongside the natural fortification aspect of the heights mentioned earlier, the eastern edge also contained the sacred Toos-nauzar (A famous champion of the Sassanian Empire [224–651], who settled in this place for a while and took it as his hunting ground), as a symbol of power

(Fig. 3.3). Its turtle-like shape reflects a long history of human presence and a distinct act of man because people revere it as a heroic site in the past life of the city. The early settlers were aware of this story and they took refuge behind the sacred hill for more security. More importantly, the orientation of the Qala (a citadel in the centre of the city on another hill opposite the Toos-nauzar hill) is towards this hill, which highlights its symbolic importance for the city. As Norberg-Schulz (1980: 5) states, "man dwells when he can orientate himself within and identify himself with an environment".

The south-west and west of the city were bordered by the Awear Mountains, forming a strong, extensive and continuous boundary to this side of the city (Fig. 3.3). This mountain range also has a sacred value due to its majestic presence and its role in refreshing the city (it has been used as a place for recreational activities, a source of drinking water also used for driving mills and a source of stone for construction purposes). It is also home to Kazzer Zendah (Peer-e gharib/invisible saint), and is known as "a place of celestial revelation for Zoroaster" (Dultshahi, 1984: 88 and 150). People usually kept the old Quran between the rocks on the top of the mountain and it was also known as Kuchick Quran, the stone of the Quran.

On the north-western side of the city, the Awear Mountain range continues and forms the Sarnuwi[1] heights. As the etymology of Sarnuwi implies, this location is at a greater height than other places as it was used for the city's water supply through a number of Qanats (subterranean canals that run in a north-west—south-east direction following the general slope of the land following the gravitational force). The north side of the city is bordered by some hills connected to the heights of Kuchka-rash (block stone) in the north-east. The break in these heights is the northern gateway of the city, which is defined by a gorge overlooking the main corridor.

The way Sanandaj is formed is a reminder of Rudolf Schwarz's quote (1949, quoted in Norberg-Schulz, 1971: 28) concerning the features of

[1] In the etymology of Sarnuwi, Sar means the higher level of something and Nuwi, according to Marufi, implies the existence of nine residential districts (villages) in the first geophysical setting of the city (Interview 27 August 2003).

landscape and its reflection in the city's structure: "The mountains are walls, the fields floors, the rivers paths, the coasts are edges and the lowest point in the mountain range, the doors".

Reflection of Natural Features on the City's Morphology

As mentioned above, the natural boundaries of Sanandaj created a bowl-like shape with a strong sense of enclosure for city residents (Fig. 3.2). This was an important defensive feature for the city of Sanandaj. As stated, the settlements along the course of the main natural corridor can be imagined as rosary beads that embrace the main city of Sanandaj (Fig. 3.2, right sketch). This can be conceived of as the order of nature which made it possible for the buildings of the city to become a point of reference for both natural features and the man-made landscape and, more importantly, its reinforcement in the features of the settlement boundaries and its form: the city walls and terraced fields and houses. This transformation of nature into a cultural landscape is what makes a space become a place with two basic properties: "Concentration and enclosure" (Norberg-Schulz, 1980: 10).

The natural boundaries and the way Sanandaj is shaped is in line with Leatherbarrow's view that "every location is really defined by physical objects" and that each element of the urban form conveys the features of its boundary in order to manifest the character of the place (Alizadeh & Habibi, 2008: 202; Leatherbarrow, 1993). This means that there is a symbolic interaction between the surrounding sites and the urban form (Kite, 2002). In the case of Sanandaj, the two fortified walls of the city (as can be seen on an 1850 map of the city), the interior wall around the Qala and the exterior one around the quarter of the Khans or political elites, are the two levels of natural barriers enhancing the defensive power of the city (Fig. 3.4). In addition, the four gates (Darwaza) of the Sanandaj city walls are the entrances giving access from the city's natural enclosure to the outside world. This is in line with Norberg-Schulz's statement that "the character of a place must be understood as a

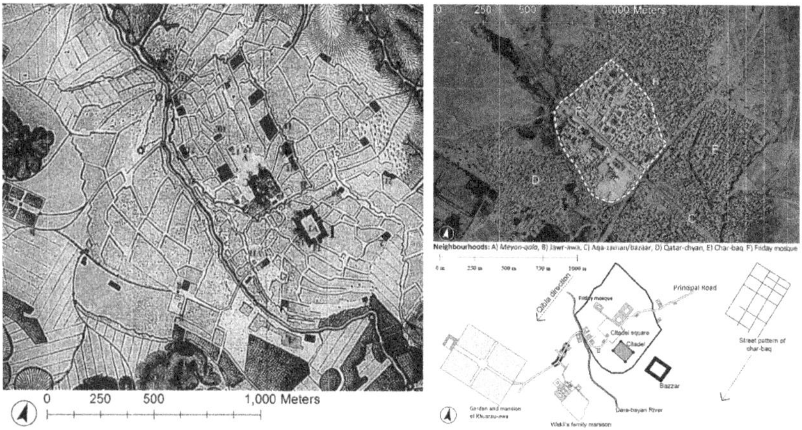

Fig. 3.4 The city's walls in 1850 and the order of natural setting in the organisation of space and social classes (*Sources* Left from Mehriar et al. [1998], right (above) from INCC (1957), right bottom by author, Autumn 2005)

product of its interaction with the surroundings" (Norberg-Schulz, 1971: 24).

Natural features of the site also shaped the social organisation of space in Sanandaj. According to Benton-Short and Short, "the city is a place where the best areas go to those with the most power and those with the least get what is left" (Benton-Short & Short, 2013: 217). Different social classes were settled according to the natural form of the city: the higher classes occupied the more desirable areas in the city centre and to the south and south-west, and the lower classes were pushed to the less desirable areas in the north and north-west. Based on the aerial photograph taken in 1957 (Fig. 3.4), the south and south-west of the city on the slope of the Awear Mountains were mainly reserved for the elites (Khans) to build their country houses to access fresh air and natural landscapes, especially during the summer time. More importantly, mosquitos usually pushed the elites and their families to move to higher altitudes close to the Awear Mountains during the summer. In world history, "poor people have often been defined as ethnically undesirable through spatial practices" (Zylstra, 2017: 7). In the case of Sanandaj, the evidence also suggests a clear geographical response to the conflict of the classes.

Furthermore, based on the Russian map (dated 1850), the bazaar contained four gates on the four sides of its rectangular form. These four gates were not exactly aligned with the cardinal points of the compass. This was due to the general topographical orientation of the city, which gave its main elements a few degrees of deviation from the cardinal points of the compass and, thus, a strong correlation with the direction of the Friday mosque (the holy direction of Qibla towards the city of Mecca). From this point, the longitudinal axis of the bazaar was oriented to the south-east towards the sacred hill of Toos-nauzar and, thus, its latitudinal axis was oriented towards the Qibla direction. These orientation techniques are also considered in the design of the Friday mosque.

The above-mentioned points can also be identified in the layout of the streets in the Char-baq (four gardens: one of the city's old neighbourhood. According to Zoroastric mystical thought, a garden is perceived as a symbolic site with streams crossing at certain points, dividing it into four separate green sectors known as Char-baq [four gardens]). Due to the general slope of the city in a south-easterly direction and the layout of the garden, the primary streets of the neighbourhood were directed towards the holy Qibla in line with the direction of the Friday mosque (Fig. 3.4). This also facilitated a logical guideline for house construction in the direction of Qibla, and the possibility of placing the main parts of the house in the northern or western parts of the courtyard and using the sun's passive energy for the harsh winter weather. This can be seen in the aerial photograph (1957), especially from the houses of the ruling class that were aligned with the direction of the Friday mosque (Fig. 3.4, aerial photograph).

Altogether, the above evidence can be conceived as a tendency in human beings to reshape the features of natural environments within the structures of cities. It means that the city takes the principles of its structure from the order of nature and that the structure of the natural acts as a foothold for humans to express their presence in the general layout of the city. It also means that the spatial layout of the traditional city is more in line with the natural setting than it is determined by cultural values. Therefore, urban forms of traditional settlements are likely to be highly "place-specific" (Rapoport, 1989: 91), which means that it is first bound up with the natural setting and then determined by social parameters (Alizadeh & Irandoost, 2017; Ozaslan, 1996). The formation of a traditional city is "an attempt to embrace location and mediate the surroundings" with the manifestation of the content of social—ultural forces, as a single entity (Leatherbarrow, 1993: 51). In other words, traditional cities work "with, not against, the forces of nature", leading to a unified urban form within its natural foothold (Kheirabadi, 1991: 39). Therefore, this interaction is a fundamental aspect of "existential space" which generally indicates "in direct and indirect ways how people have related to the nature and what kind of existence they have created there" in order to embrace the outside world, and "bring the world close" (Davis, 1997: 140).

City Structure

After understanding the geophysical settings and some points concerning the reflection of the natural features in the overall morphology of the city, the structure of the city's built form and its constituent elements are analysed in this section, to explain how the details of place aggregate to each other. At first glance, the city is divided into two halves by the river of Dara-bayan running in a west-easterly direction, according to the Russian map of 1850 and the aerial photograph of 1957. The main sector of the city is located on the northern side of the river comprising three neighbourhoods as well as the bazaar complex, the Friday mosque (Jamia), the Qala/citadel, the ruling elite quarter and the city walls (Fig. 3.4). The southern sector includes the neighbourhood of Qatarchyan and the mansion and garden complex of Khusrau-awa. Generally, the aggregation of these components implies a semi-planned structure, the main elements of which are organised in a rectangular form, indicating the notion of formal planning and the layout of the neighbourhoods' spatial units being analogous to the organic form (Fig. 3.4). Applying these two concepts highlights the notion of defensive purposes, as the main parts, the Qala, the houses of the ruling elite, the bazaar complex and the Jamia mosque were surrounded by the compact form of the neighbourhoods. More importantly, within this general pattern of aggregation, the two elements of the Qala (the seat of power) and the bazaar complex (the merchants' setting) were highlighted due to their major role in the city life.

City Walls

As was made clear in the historical background, Sanandaj possessed a peculiar structure. It included a central fortified Qala within the fortified inner city (the quarter of nobles) and the quarters of commoners, which was not surrounded by wall (Fig. 3.4). Considering the concept of multiple edges conceived by Kostof (1992)—the castellated palace (citadel) and walled suburb occur in every culture—this type of embattled wall can be regarded as a different projection of edgedness. The main sector of the city was located behind the city's walls with no man-made edgedness except the edge of a compact group of settlements and nature. Such perception of the third layer of edgedness was witnessed by Rich (1836: 207), who visited the city in 1820. On his first glance at the city from the *Qala*, he perceived it as being walled by three edges: "1) the castellated palace, situated on the high mountain, where he had stood, and was surrounded by a wall; 2) at the foot was another part of the town and another embattled wall; 3) beyond this wall was more of the city which again was surrounded by the wall". But when he left the city and saw it again from the suburb, he realised that the third part was free of man-made walls. It was only the compact form of the settlements that had convinced him to consider the third layer of edgedness for the city as a whole.

Such a pattern of city walls can mainly be observed for the pre-Islamic period, when the idea of class division was important and its value was reflected in the urban space. During that period, the city comprised three districts: (1) The dej/citadel, including the palaces situated in the highest area of the city, which was surrounded by a wall. (2) The inner-shar/city containing the houses of nobles, which again was walled. (3) Beyond these two structures, the outer Shar comprised the residential quarters, bazaar, gardens and suburban areas (Habibi, 1996; Sultanzadeh, 1986).

The pattern of the city walls can be traced back to the frontier position between the Ottoman and Safavid Empires. The city was founded with a definite purpose, "namely as a politico-administrative centre of Kurdistan" (Vasilyeva, 2000: 5) for the Safavid empire. More than four centuries' experience of mound-towns preceding the foundation of the city (1168–1639) formed the basic nucleus of Khan's authority (as the representative semi-feudal community) in the region. The construction of a city usually began with a strong-walled citadel/Qala on the top of a hill commanding the domain and the settlements at the foot of the hill. Therefore, the city walls were not just boundaries, but rather an element of defence and the symbol of power and class distinction. In that period, convincing evidence of class stratification can be found in the words of Vasilyeva (2000: 8) who claimed that "the wealth in itself was not an indicator of the social status" because "the eldest son of Amanullah-khan (1800–1825) was not chosen as his successor due to low background of his mother who was the daughter of a money-changer/*Seraf*" (Rich, 1836: 209).

Physical Features of the City Walls

Based on the Russian map dated 1850, (Mehriar et al., 1998), it can be concluded that the city walls were more or less rectangular in form. It was oriented according to the cardinal points of the compass with some degrees of deviation due to the general slope of the city's location, which set its longitudinal axis in the north-west–south-east direction (Fig. 3.4). Its partial deviation from a rectangular form was also due to the effects of the slope and the course of the Dara-bayan River on the southern side of the walls. The second wall, which enclosed the elite quarter, was linked to the other quarters through four fortified gates on its four sides.

The gates of Iranian cities generally were mainly situated in the direction of the trade routes to control the flow of people, animals and goods from the suburban areas. In contrast, the gates of Sanandaj were built in the interests of the ruling class and in accordance with their needs. They served as threshold to the world of ruling elites to sift the flows of people from other quarters, a clear symbol of the Khans' socio-political

Fig. 3.5 The seat of power on the top of a hill in the centre of the city in the late period of Naser o-Din Shah, 1848–1896. The image of Khusrawia with pillars in front is at the right corner of the hill and the Jamia Mosque is clear in the right corner of the photograph (*Source* From collection of Gazarani, date unknown)

status. For this reason, each gate faced one of the quarters. The northern gate was directed towards the quarter of Jawr-awa, the servants' quarter, named the gate of Tapula. The eastern gate faced the bazaar complex and the merchants' quarter, titled the gate of Agha-Rahim. The southern gate was constructed to link the ruling class to the slopes of the Awear Mountains, the site of the garden complex of Khusrau-awa and the quarter of *Qatar-chyan*. It was called the gate of Abull-azim. Finally, the western gate—the gate of the Dareh/valley—faced towards the source of water in the Dasht-e Sarnwi, from where the networks of Qanats (subterranean canals) originated for the water supply of the city's main mansions (Fig. 3.4).

Residential Neighbourhoods (Mahallaha)

As explained, Sanandaj was formed and evolved from the concept of mound-towns, which established the walls around the palaces and the ruling houses as the symbol of the *Khan's* authority. This diverges from the general pattern of neighbourhood formation in Iranian cities, which was usually bound to the linear form of the bazaar as the backbone of the city. This typical form of "self-centred" and "self-contained" development corresponds to the interests of the *Khans*. To some extent, the centre of gravity of the city was based upon the source of power (*Qala*) instead of on the bazaar, because its political importance far outweighed its economic function. Therefore, principal road followed a course along the inner-city, which was the centre of gravity. This road leads from the north-east (where a parade of troops usually accompanied the new Wali and other Iranian authorities to the inner-shar and then to the square in front of the Qala) to the south-east (where the summer house of the Wali was located on the slope of the Awear Mountains). Most of the neighbourhood evolved along this road and the bazaar was excluded and maintained as a point (instead of as a line) within the maze pattern of streets outside of the city walls. This orientation made the neighbourhoods centripetal cells of the city to such an extent that Vasilyeva (2000: 7) stresses:

> The [neighbourhoods] mahallas were the basic structural and territorial cells of the city and the basic units of the town society. They are regarded as village-like communities within the urban whole, as [some of them even] resembled physically separated entities.

In other words, the neighbourhoods were organised as a collection of villages, which circled the nucleus of power in the form of satellite quarters according to their social and professional backgrounds (Fig. 4.3). This divided the city into four main quarters, so that the central part was occupied by the seat of power and ruling house (*Meyan-qala*) which formed the inner-city. The outer city comprised Jawr-awa to the north as the servants' neighbourhood, Mahalla-*ye* bazaar (Aqa-zaman) in the east as the location of the merchants (Bazaaries) and Mahalla-ye Qatarchyan in the south as the place of muleteers (who transported the goods). These four neighbourhoods embraced the city centre. It served to supply the needs of the ruling class and provided the services that a medium-sized city needed to function. Each Mahalla was in turn divided into subsections, which were also called Mahalla. Generally speaking, such focused power, along with the geophysical setting of the city, highlights the concept of a concentrated centre rather than that of a linear development.

This organisation of the city space to some extent conveys the structure of the Iranian city in the pre-Islamic period. As Falamaki (1987) points out, the structure of the old Iranian city was directly dependent on the governmental-administrative order, insomuch as the form of its physical space followed this order. In this spatial order, the citizens were divided into four classes due to their professional backgrounds, and each was allocated a special status and delimited position. This means that there was a kind of coexistence structure, which sustained the life of the traditional city. "The city helps to sustain the ruling elite, and in turn the city's own health, its growth and survival, requires a well-developed power structure" (Kostof, 1992: 74).

Before moving on to the description of each neighbourhood, one may ask why we do not describe the organisation of the city's neighbourhoods according to tribal attributes, as did Barth (1953), McDowall (1996), and Ross and Mohammadpur (2018), who frequently looked upon Kurdistan from an angle of tribalism.

As explained, Sanandaj is not an organic city that evolved from an existing rural settlement over the course of time, within which a tribal structure could gradually emerge. It is a founded city, brought into existence in 1639, as a result of a peace treaty between two rival empires. Its population came mainly from four ruined Qalas/citadels in different regions of Kurdistan within the Ardalans' territory. Thus, the city's population came from different social backgrounds far from any tribal confederation. In addition, due to the nature of tribalism in Kurdistan, where the political dimension outweighed blood ties as it mostly implied a "defensive organisation" (Barth, 1953; Nikitine 1983), it was not in the interests of the Wali and ruling class to allow the formation of tribal organisation in the city. The evidence of this point can be traced in the words of Rich (1936), who observed a discussion between two rival families, Baban and Ardalan. "The Wali of Ardalan once asked Abd al Rahman Baban why his own servants, though generously treated, would never follow him into exile nor show any personal loyalty, such as Baban retainers usually demonstrated" (McDowall, 2001: 91).

> The answer of old Abdurrahman Pasha was very characteristic. 'You are not,' said the old chieftain, 'the lord of a tribe, nor are your men your tribesmen. You may clothe, feed them, and make them rich, but they are not your cousins; they are but your servants!' (Rich, 1836: 86)

More importantly, tribal associations mostly existed among the nomadic Kurds who had a mobile structure between the two empires. Thus, within the walls of the cities, tribal affiliations were weakened (Izady, 1992). This point does not indicate the existence of a hard line between the city and the countryside, especially the nomadic life. As Nikitine (1983) indicates, the feudal master (such as Wali Ardalan) not only imposed his power on settled people, but also on the tribes, especially to provide forces in the times of war against rivals.

> The riders of tribal origin participated in all military actions undertaken by the Ardalans. The tribe provided the city with military energy and power and nomad camps served as a shelter for the representatives of the prince's house in danger. (Vasilyeva, 2000: 12)

Rich (1836: 217) explains that, although the Wali of Ardalan unlike his Baban colleague (in the Iraqi region) did not act as the head of tribes, "many nomadic Kurds were under his jurisdiction, which were integrated in occasion of war against hostiles". These were branches of the Sheikh Ismaeli, the Merdemeh, the Kelhor and Jaf tribes. The agricultural products from this way of life (a mobile pattern on the frontier of the two empires) formed the backbone of the economy as the product from Sanandaj most in demand was butter, which was exported to other regions and even to the court of the Safavid Empire (Vasilyeva, 2000).

The above points again reinforce the city's social structure based on the professions—each fulfilling a particular function. This point is more evident when the names of quarters and their subdivisions are considered. Their etymology mostly reflects the professional backgrounds of the inhabitants and there is no link with a particular family or tribe. The importance of this factor was such that Vasilyeva (2000: 9) states, "the establishment and growth of the city was related to the distinction of professionals".

Meyan-Qala/the Ruling-Class Quarter

As the name of Meyan-qala in Kurdish signifies, the citadel in-between, it comprises the central part of the city, which contains the seat of power and the houses of the ruling class. The seat of power was located on the higher part of the site in the south-eastern corner of the quarter, overlooking its surroundings and surrounded by its own battlement wall to secure greater privacy (Fig. 3.5). An impression of this castellated structure can be found in the words of Kinneir (1813: 144, 145), a visitor to the city in 1810:

> The Wallee, who seldom quits this place, resides in a sumptuous palace, built on the top of a small hill in the centre of the town, where he maintains a degree of state and splendour superior to anything I have seen in Persia, except at court.

According to Gaube (1979: 22), the location of the government's court at the highest point of the settlement is one of the characteristics of early Iranian cities in general. Izady (1992: 263) recognised it as a basic design of cities in Kurdish culture, which was adopted into the architecture of other cultures.

Due to the mass destruction of the site, especially after the removal of the Ardalans in 1867, the detail of its constituent elements is not very clear and needs more archaeological examination. The only exception to this point is Rich's observations, which make clear some of the details. He noted that the site of power contained the following buildings: a pavilion called Khusrawia, an old audience chamber supported by pillars at its front, built in the period of Khusrau-khan in 1773, an alcove at the back of the hall that was named Shah-nesheen/Talar, a new splendid audience chamber larger than the Talar, built by the then Wali which contained various portraits of Iranian and non-Iranian kings, two beautifully ornamented galleries on both the sides of the new hall named Bala-khane, and many other buildings under construction. More recent studies concerning the Qala reveals that its site contained a geometrical square form; "a concentrated form of centre flanked by four Talars oriented to the cardinal points of the campus" (Pars, 2004: 84).

The inner-city also accommodated the ruling houses, those of seventeen noble families—"the bearer of the political fate of the region, namely the politico-administrative centre" or ruling class (Vasilyeva, 2000: 5, 7). As explained, it was completely separated from the city of commoners by the fortified wall and gates as the symbol of power and class distinction. This conforms with the view of Kostof (1992: 75, 76), who says that there were two primary impulses that motivated the formation of the "sovereign district" in the centre of a town with fortified walls—"the dignity of the ruler, and his safety against the rival provinces".

Apart from the image of the walls, the size of houses with the grand scale of the gateways and central courtyards, mainly imitated from the Safavid buildings in Isfahan (Rich, 1836), also signifies the distinctive character of this quarter (Fig. 3.6). In other words, they communicate the notion of wealth and power. It was a self-reliant quarter, equipped with the common elements of an Iranian city: mosque, Hammam (bath-house), Bazaarche (small market) and public spaces. Besides, it also

Fig. 3.6 An impression of the courtyard structure, the grandeur of the gateways to the houses and the size of the houses in the Quarter of the ruling class. (A) The ruling-class quarter/inner-city. (B) The neighbourhood of Jawr-Awa. (F) The Jamia mosque (*Source* Aerial photo from INCC, 1957, above photo by Mr. Kasra Ketaboullahi, autumn 2020, and bottom photo by author, autumn 2003)

contained the Jamia mosque and the urban element of a square in front of the Qala facing to the west (Fig. 3.4).

The street pattern was irregular and to some extent conditioned by the topography and the social background of the inhabitants. As it was home to wealthy people, the size of houses was relatively large, which resulted in large blocks with relatively straight streets. The main streets towards the south, bordered by the river, follow the contour lines of the slope. It was the only quarter supplied with a sewer system that ended in the Dara-bayan River. Generally, due to the extent of the physical changes that took place in the course of time in this quarter, the street pattern is not clear enough for further investigation.

Jawr-Awa Neighbourhood

As one of the neighbourhoods of the city of commoners, it was located at the north side of the inner-shar (the quarter of Meyan-qala) named Mahalla-ye Jawr-awa (Fig. 3.4 and 3.7). As the etymology of its name indicates, it was the place of toilers, that is, the city's servants and workers. These people existed due to the nature of the "quasi-feudal" (McDowall, 2001) structure of the community, where the Khan possessed high rank and status in the community and thus imposed his power over the peasants and also, partly, over the nomads. As Lambton (1991: 140) indicates, the land-owning classes were the most powerful elements especially in the Qajar kingdom, which formed the concept of the Khan, reinforcing their role as a link between the state and the area under control (Garthwaite, 1981). Bruinessen (1978: 81, 82) describes Xan/Khan and Beg as originally feudal titles, referring to people who were literate, civilised and engaged in politics. They often possessed large land-holdings in the plain, which were cultivated by non-tribal peasants.

In the semi-feudal structure of the community, the life of the peasants was tightly bound to the land, so that they had to work for the benefit of the Khans. From this point, "the peasants, moreover, have little or no security from their Khans, who tend to look upon them as chattels" (Lambton, 1991: 291). This relationship between the Khan and the peasants continued in the city in the form of the ruling class and its servants.

It meant that the ruling class remained the main property owners of the region and each possessed some of the villages, controlled by representatives called either Kadkhuda or Zabit, who also collected the Khan's share of the agricultural products. Thus, they were referred to as absentee landlords (Bruinessen, 1978). From these points, it can be inferred that these people were mostly the poor from the countryside, who were moved to the city by the ruling class to work as their servants.

Considering the above points and due to their low status, these former peasants were mainly settled in the northern half of the city, away from the city centre, where the hilly land of the site posed the problem of water supply as well. The general morphology of the neighbourhood, particularly the street design, was mainly affected by the physiography. Apart from being a neighbourhood of servants, a part of the social structure of the city in general, one can observe differences relating to wealth and status especially in the areas neighbouring the ruling-class quarter. The morphology of the houses close to the ruling-class district is completely different from that of the houses located further from that district. Close to the border between the two neighbourhoods, one would thus observe major changes in the form, size and spatial structure of the houses on moving further from this line towards the north. In other words, it means that the families holding a higher status settled closer to the quarter of the ruling class and those of a lower status moved further away, onto the hilly area of the neighbourhood (Figs. 3.6 and 3.7).

The street grid follows no specific pattern, because the general layout of the streets was affected by the physiography of the site, so that the main streets followed the natural gullies leading from the northern heights of the city site. Proximity to these gullies led to the formation of two primary streets in the neighbourhood (Fig. 3.7). Apart from the social status of the inhabitants, this can be recognised as the effect of the natural setting, which compelled the inhabitants to protect their houses from unexpected flood risks. Parallel to these streets, another was formed along the existing man-made route leading towards the countryside. These three streets made up the backbone of the neighbourhood as they crossed the quarter and provided a link to the principal route. The other primary streets (Kuche) branched out from there to make connections in-between and give access to the houses. This means that

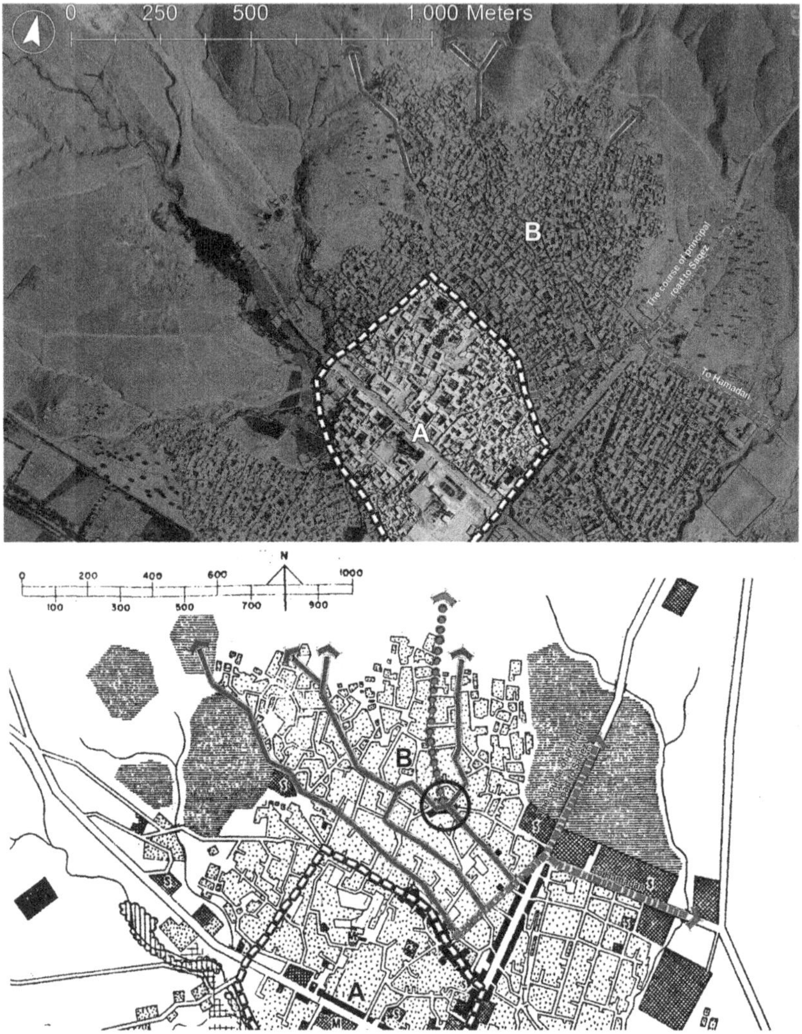

Fig. 3.7 The role of topography in the layout of the street pattern in the neighbourhood of Jawr-awa. (A) Ruling class, (B) Jawr-awa (*Source* Aerial photograph from INCC, 1957 and the map from Alten [1958])

the movement of people was channelled from those three streets to the principal route leading to the north gate of the inner-shar or to the site of the bazaar.

Such a pattern of movement above is markedly different from the common pattern in other Iranian cities, where the main thoroughfares (Raste) are usually linked to the linear form of the bazaar, which crossed the city and formed its backbone (Kheirabadi, 1991; Sultanzadeh, 1986). Here, each Raste is linked to the principal road leading towards the Qala instead of the bazaar. As explained before, the reasons for such peculiarity relate to the general socio-spatial factors which caused the built form of the city to be mainly in line with the gravity of political powers located within the inner-shar, and not with economic powers. Thus, the bazaar was formed in a rigid structure within a compact form of settlement, away from the principal route. Within the circulation system of the neighbourhood, the centre of activities called Bazaarche Istanbuli was formed along with one of the thoroughfares, the existing man-made road to the countryside, where it contained the general public buildings of Iranian cities comprising a mosque, Hammam/bathhouse, chaikhana/Qahva-khane and some retail shops. It did however not contain the spatial element of Hussainiya, specific to Shiite ideology, due to the religious background of the inhabitants who adhere to the Sunnite ideology.

Aqa-Zaman/bazaar Neighbourhood

This neighbourhood was located to the east of the inner-shar, and called Mahalla-ye bazaar or Aqa-zaman (Fig. 3.8). This was the merchants' district, which encircled the rectangular form of the bazaar. It originated mainly due to the function of the bazaar and thus evolved away from the line of the city's principal road.

As discussed in the historical background, the city population also comprised religious minorities of Jews and Armenians, who were moved in from the northern region of Iraq and some from the villages in the region to manage the economic activities of the city (Ayazi, 1992). Due to their professional background and well-known reputation as traders

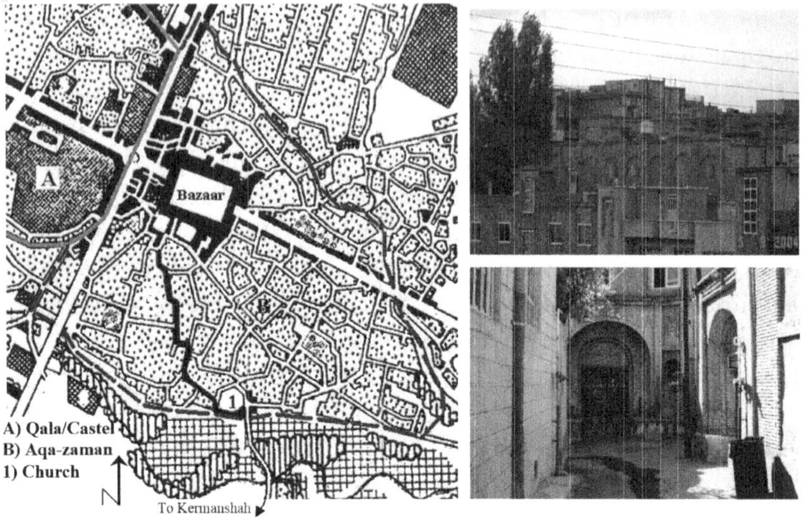

Fig. 3.8 The maze pattern of streets in the neighbourhood of city's minorities/Aqa-Zaman and the Church entrance and its external façade, close and far view. The street pattern of neighbourhood suggests the notion of defence. The map also shows how the urban element of the bazaar as the focal point of the neighbourhood polarised the streets into a more defensive pattern. The crossed-shaped streets were laid out by French engineers in 1938–1939 (*Source* Map from Alten [1958] and photos by Author, August 2003–2004)

(Nikitine, 1983), they mostly settled in this quarter close to the bazaar. Kinneir (1973), who visited the city in 1810, reports that they mostly traded with Mosul, Baghdad and Isfahan. Each of these two religious minorities occupied a particular part within the neighbourhood. The Armenians lived in the southern part, called Mahalla-ye Masihyan, and the Jewish inhabitants lived in the northern part, called Mahalla-ye Kalemiha. Each had their own place of worshi. As the Jewish minority was more numerous than the Christians, it had three synagogues. One was called the great synagogue and was used for special ceremonies. None of them bore any special architectural feature or symbol, so that they are not recognisable among other residential houses, because the Jewish minority used to convert a residential house into their place of worshi. This was also noted by English (1966) in the city of Kerman. In contrast, the rectangular form of the city's church was oriented according to the cardinal points of the campus, so that its longitudinal axis is in the eastward direction and the interior contained features of the church. Although its exterior facade was considered to be in harmony with its context, it is a fine grain urban element in the neighbourhood recognisable by the large scale of its walls and the small domes on the roof (Fig. 3.8).

Following the socio-political changes that took place after the reign of the Qajar dynasty, over time, most of the Armenian people migrated to the city of Tehran or other major cities in Iran, and the Jewish people largely moved to Israel. Today, only a few families adhering to these religions remain in the city, but the physical character of the neighbourhood, with their places of worship, still recall and communicate the life and memories of those minorities.

Being minorities and engaging in trade activities with the symbol of wealth were strong inducements for these groups to contextualise a particular layout for their neighbourhood, differing from that of other neighbourhoods. The pattern of streets strongly conveys a high concern with security, rather regarding the invasions of hostile tribes from the countryside than the inhabitants of other quarters. As Ayazi (1992) indicates, they lived in an atmosphere of friendship with other people, indeed, to such an extent that Muslim people also lived in their neighbourhood and they had a good mutual relationship, even attending each other's ceremonies.

The neighbourhood layout referred to is the complexity of the street pattern, which can be recognised as an "original vernacular solution" to the need for defence (Norberg-Schulz 1980: 118). This pattern was conditioned by the structure of the residential blocks, which is similar to the shape of the cells of a beehive. The streets are narrow, and continuously change their direction (Fig. 3.8). More importantly, the network of streets close to the bazaar formed some semi-circles, which blurred any direct line of access to the bazaar site. Such a twisted street pattern enhanced the defensive aspect of the neighbourhood in general and the bazaar in particular. In the viewpoint of Kheirabadi (1991: 34), the maze form of "the alleyways served to disorient strangers and to hinder the manoeuvrability of plundering horse-men in the confined quarter". Similarly, in the words of local historian Marufi,[2] these narrow and twisted alleys usually limited the flow of hostile tribes to smaller groups, and gave the inhabitants the opportunity to defend themselves. It must be also emphasised that the neighbourhood displayed a minimum cul-de-sac structure and a more circular form than the other neighbourhoods to confuse the movements of invaders (Fig. 3.8). This is what English (1966) also observed in the street pattern of the Jewish neighbourhood in Kerman. He describes the lanes of the Jewish quarter of Kerman as extremely narrow and even more irregular than that of the Muslim residential neighbourhoods.

Although the neighbourhood was characterised by the bazaar complex, it was also supplied by its own Bazaarche in the vicinity of the church. Here the church acted as a focal point for other urban elements such as the Hammam and retail shops. Based on the Russian map in 1850 (Mehriar, 1998), this centre also contained the element of a square in front of the church. The existence of the church square, as it is not seen in other Iranian cities, to some extent recalls the same concept in the medieval towns of Europe, especially "the burial ground" or Parvis (Morris, 1994: 76). The square was situated along the road from the countryside, south of the city. Presumably it was linked to the caravan road from the city of Kermanshah, because the extension of the road—with some breaks—leads to the caravansaries of the bazaar (Fig. 3.8). This square was called the square of Kilisa, or Church Square.

Qatar-Chyan Neighbourhood

The neighbourhood of Qatar-chyan and the garden complex of Khusrau-awa on the southern side of the inner-city comprised the fourth section of the city. It was separated from the other neighbourhoods by the Dara-bayan River (Fig. 3.4). As explained earlier, this neighbourhood was the place of muleteers who usually transported the city and countryside products to other cities, including Hamadan, Zanjan, Tehran, Isfahan, Rasht and to some parts of Iraq, namely Mosul, Sulaimani, and Baghdad.

[2] Interviewed 29 August 2003.

On their return, they also supplied the demands of the city merchants as well as the necessary goods for the city and its region. Because it was well known and because it was separated from the inner-city and other neighbourhoods, it contained two active Bazaarche and another seat of power (Wakil's edifice). Because an analysis of its local structure will follow later, we have here simply mentioned a number of general points.

Char-Baq Neighbourhood

For a long time, the main structure of the city consisted of the four large neighbourhoods mentioned above. This lasted until the mid-nineteenth century, when a new area, with a peculiar physical structure, was added to the city. It was the neighbourhood of Char-baq which lies to the north-east of the city (Fig. 3.4). As the name Char-baq (Four Gardens) conveys, it was built on the site of previous gardens, divided by crossroads named Baq-e Mayan. The structure of the garden became a basic ground plan for the street grid of the quarter and had a rectangular form, distinguishable from the street pattern of other neighbourhoods.

Based on the local chronicles, the creation of this neighbourhood can be dated back to the nineteenth century when a new Char-baq was created in the period of Amanullah-khan the great (1799–1820) and favoured by the Wali, thus causing the deterioration of the former Char-baq (Sharafkandi, 2002; Vasilyeva, 2000). This point, along with the natural growth of the city population, especially those who engaged with the function of the bazaar, led to the occupation of the site and its gradual construction as another neighbourhood. Compared to other neighbourhoods, especially Jawr-awa and Qatar-chyan, the physical development of the bazaar neighbourhood was limited by natural boundaries comprising the Dara-bayan River to the south and south-east, and gullies to the north and north-east. Thus, the site of Baq-e Mayan, which had deteriorated, was the site closest to the neighbourhood to be occupied.

From the above points and considering the retained rectangular structure of the garden in the layout of the streets (Figs. 3.4 and 3.9), it can be conceived that wealthy people settled there. The convincing evidence

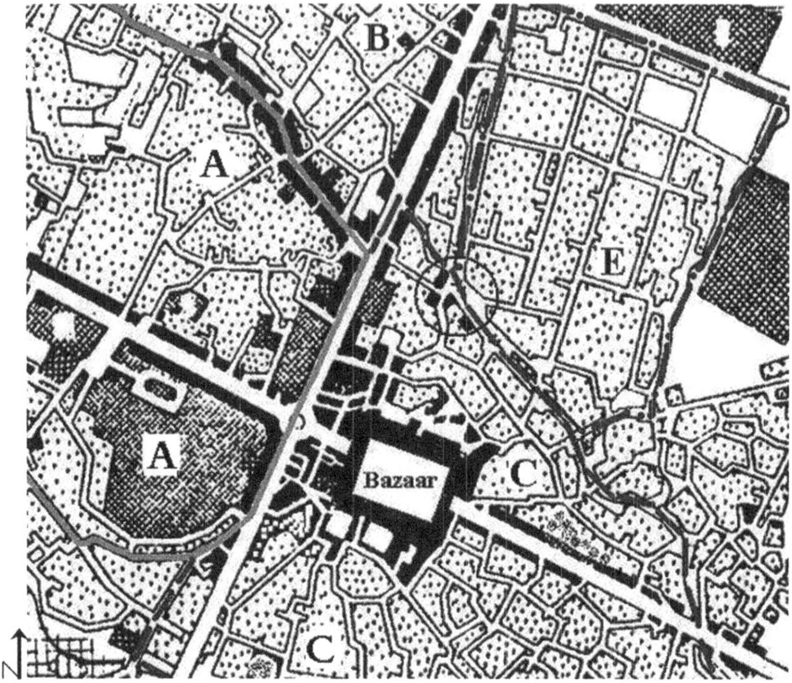

A) Ruling class, B) Jawr-awa, C) Aqa-zaman/Bazaar, E) Char-baq

Fig. 3.9 The grid of streets in the neighbourhood of Char-bag. This pattern was retained from the garden structure of Baq-e Mayan and displays its contrast with other neighbourhoods (*Source* Alten [1958])

is the relative size of houses, which are larger than in other neighbourhoods, except the quarter of the ruling class. Although some of the alleys were structured according to a Bunbast (cul-de-sac) pattern, this is much rather a recent spatial feature arising from plot divisions between later generations. More importantly, most of the Bunbasts contained just two or three small houses.

Due to the general gradient of the city site and the initial layout of the garden, the primary streets of the quarter run in the direction of holy Qibla (Kaaba), in line with the orientation of the Jamia mosque. Thus, the secondary streets branched off at right angles to the primary streets, providing access towards residences. This provided a logical guideline for house construction in the direction of Qibla.

The element of centrality is not very clear or strong in this neighbourhood, due to its uniform structure, the rectangular pattern of streets, and to the fact that it was outside the main structure of the city. The satellite structure of other neighbourhoods circled the inner-shar quarter and thus gave them the possibility of being self-contained and self-centred structures. Based on the existing structure, it is presumed that the centre was in the intersection of streets adjacent to the neighbourhood of Agazaman/bazaar. It contains the urban elements of bath house, mosque and retail shops.

Public Places

After recognising the general structure of the city, the main public spaces, which played an important role in the social life of the city, are examined in this section. Considering the relevant literature (Amin, 2008; Gehl, 2011; Holland et al., 2007; Madanipour, 2013; Oldenburg, 1989) public spaces, as a part of the public sphere, can be defined as a free meeting space or gathering places outside the home and workplace that are generally accessible by members of the public. At the same time, they cater for the daily needs of households and stimulate optional activities, thus fostering spontaneous meetings affecting people's interactions and opportunities for proximity and contact. Regardless of some ambiguities in defining the domain of public places in Islamic cities, the realm of this zone can be recognised "outside the boundaries of individual or small group control, mediating between private spaces and used for a variety of often overlapping functional and symbolic purposes" (Ardalan, 1980b; Madanipour, 2003: 112, 113). In the words of Ardalan (1980b: 5), this, in traditional Islamic culture, would include "the public garden, passages, streets, covered streets such as bazaars and Suqs and other traditional places of commerce, as well as places of the institutions of the society together with their urban contexts". In this definition, the public space of the Maidan/Square was disregarded, the element which was recognised as the node and the focal point of socio-political[3] and

[3] "The two revolutions that have shaped Iran's modern history were both performed in public spaces of the Baharestan square in front of the parliament in Tehran" (Madanipour, 1998: 209).

economic activities of citizens. For Kostof (1992: 127), in this regard, who believed that there is a well-defined sense of public space in Islamic cities, the main domain of public spaces can be divided into "streets [covered streets, bazaar], Maidans (squares), mosques, and cemeteries". From one perspective, both definitions can be criticised as ignoring the three-fold organisation of space in Islamic cities, which distinguishes intermediary spaces at levels between the two basic realms of public and private domains. This comprised the semi-public and semi-private spaces within the circulation system of the streets, mainly those places that define women's social realm. Generally, through combining points of both notions as mentioned, we can illustrate the domain of public places as follows: Maidans, religious centres (comprising the Jamia mosque, mosques in the quarters, the Takaya and Mazars/shrines and cemeteries), streets, bazaars (both the main covered bazaar and the open ones in the neighborhoods), local institutions and public gardens. Within this section, the public places of Maidan, main bazaar, Jamia mosque and principal road are discussed and the other elements will be dealt later on alongside the details of place structure concerning the neighbourhood of Qatar-chyan.

Maidans (Squares)

Based on the Russian map (Mehriar et al., 1998), the city contained three Maidans in the period of the Qajar dynasty including Maidan-e Wali or Arg in front of the fortified wall of the Qala in the quarter of ruling class, Maidan-e Alaf-khan and that of the Kilisa/Church in the neighbourhood of Aqa-zaman/bazaar (Figs. 3.4 and 3.8). In the following sections, each of these three squares is discussed in detail concerning their socio-spatial characteristics.

Maidan-e Wali (Arg square)

As Sultanzadeh (1986) and Bakhtiar (2011) indicate, most of the historical books concerning the Iranian city examine the formation and evolution of the Maidan in relation to the function of the government or court where troops were usually paraded in front of imperial palaces. It means that the concept of the Maidan was initially bound to the affairs of government and its related ceremonies. In later periods, other socio-economic activities were embedded into the function of the Maidan. In the view of Habibi (1996), the element of the Maidan was contextualised in Iranian city since the period of Safavid dynasty onwards, to the extent that each human settlement (city or village) was characterised in respect of a Maidan. He states that the urban element of the Maidan was the centre of gravity for governmental, religious and economic activities and in some cases accommodated mixed functions. In this regard, he referred to the city of Isfahan, which "boasted a well-developed inner core" centred on a great Maidan (Bakhtiar, 2011: 150) named Naqsh-e Jahan (the pattern of the world) as the "forecourt to the imperial palace and mosque, open-air market, and polo field" (Kostof, 1992: 131).

As the political importance of the city far outweighed its economic function, the first public space of the city can be recognised as being the open space in front of the fortified wall of the Qala, named Arg sqaure. It was the place of military activities and execution of sentences, where the rulers paraded their troops for war and official ceremonies. Kheirabadi (1991: 77) links the Arg square to "the cities with significant strategic importance in the region where the Arg served as a citadel for regular practice of ruler's armies". Sultanzadeh (1983) ascribes it to the major Iranian cities, due to the existence of a governmental Arg, which communicates power. For example, he refers to the city of Qazvin, which contained the element of a Maidan in front of the seat of government in the period during which the city was the capital of Iran (in the sixteenth century under the Safavid Tahmasb I). Considering the symbolic function of the city's walls, which separated the seat of power from the commoners, the viewpoint of Madanipour (2003: 204), is also logical, when he says that "the political significance of public space may have been manipulated in cities where the rulers have kept an often violent distance from the ruled".

As a basic principle of the Iranian city, the spatial element of the Maidan was usually found at the meeting point of main streets or by aggregating the main architectural elements of the city. In line with this principle, Maidan-e Wali was formed along the principal road running from the north to the slopes of the Awear Mountains to the south, at its junction with the Kuche Deraz (long alley) leading towards the Jamia mosque in the west (Fig. 3.4, 3.10 and 3.14). This structure, in correlation with the Maidan as a centre and streets as an axis, reflexes Norberg-Schulz's (1980: 58) idea that

> When an opening is introduced in a centralized enclosure, an axis is created which implies longitudinal movement.

The location of the Wali square within the inner-shar and its spatial relationship with the main bazaar manifests an important contrast with other Iranian cities in the periods of the Safavid and Qajar dynasties. There, the element of Maidan-e Arg was usually in direct connection with the main public places of the city (the bazaar and Jamia mosque), such as Sabzeh-maidan in Tehran and Naqsh-e Jahan in Isfahan or Maidan-e Ganj-Ali-khan in Kerman. However, here the element of the Maidan was created closely in line with the function of the Qala, so that the Jamia mosque and the main bazaar were set away from it. Although the previous Jamia mosque, on the slope of the Qala hill towards the east, was destroyed in the period of Amanullah-Khan the great (1799–1825), the new one was not erected in the vicinity of the Maidan. It was, rather, detached from the site of the Qala and the main bazaar for some political and religious purposes. The reason for this will be discussed later. These notions define the Maidan-e Wali as a place solely for displaying power relations, without any other socio-economical meanings similar to those of Maidans in other Iranian cities.

In addition to the above points, Maidan-e Wali was surrounded by the spatial elements of the Peer-Omar shrine and Naghareh-khaneh[4] (the place where the drums were beaten at sunset). The square was situated along the principal road running from the north gate of the inner-shar—the gate of Sartapula. Here, the principal road channelled the flow of troops and other official welcoming ceremonies towards the city walls. Presumably, the procession to commemorate Muharam moved through

[4] As Tavassoli and Bonyadi (1992: 25) indicate, the function of Naghareh-khaneh was that of announcing the time of sunrise and sunset.

this road to the city wall and then to the Maidan-e Wali where the shrine of Peer-Omar abutted the square. The convincing evidence for this is that the shrine of Hajara-khaton and Amjad's house (the only place used by Shiite people as Husayniyah) are adjacent to the north gate. Further evidence is provided by the presence of some Shiite people in the quarter of the ruling class. The starting point of this commemoration usually was in the Amjad's house and Hajara-khaton shrine. From there, the people moved towards Maidan-e Wali through the northern gate (Fig. 3.10).

In comparison with the functions of other squares in Iranian cities—the three squares mentioned earlier in the cities of Isfahan, Tehran and Kerman—and notwithstanding the commemoration of Muharam, it can be conceived that Maidan-e Wali was not a public place intended for use by all people. As it was situated in the quarter of the ruling class with a fortified wall, it can however be recognised as a semi-public place in the city, a place that served to communicate the power focused on the Wali, and that functioned as a meeting place for the ruling class. For this reason, it can be defined as a place representing political life in the city.

1) Arg square, 2) Peer-Omar, 3) Hajara-khaton, 4) Amjad's house, 5) Jamia moque

Fig. 3.10 The spatial relationship between Maidan-e Wali and the shrines of Hajara-Khaton and Peer-Omar. The right side are two old photographs of the Maidan-e Wali/Arg square were taken in the late period of Naser o-Din Shah, 1848–1896. In the above photo the Jamia mosque is clearly seen behind the Maidan. The down photo shows the entrance gateway to the court of Wali facing the Maidan and people's assembly in front of the gate (*Source* Map by Author, Autumn 2005, Photos from collection of Gazarani, date unknown)

Bazaar and Maidan-e Alaf-khan

As the second Maidan was formed by the square structure of the main bazaar, it is examined with the discussion concerning the public urban element of the bazaar. The bazaar can be recognised as a common feature of public space in different cultures. In the Iranian context, the concept was born in the Achaemenid Empire (550–330 BCE). It further evolved during the Sassanid Empire (224–651 CE) and, lastly, it gained its final form in the period of the Safavid dynasty—as the main city's spine connecting prominent architectural elements and public places and forming the city neighbourhoods along its path (Pourjafar et al., 2014; Sultanzadeh, 2001). Its manifestation in the city of Isfahan was recognised as "one of the great spatial sequences of the world" (Browne, 1976: 260).

Literature mostly recognises the structure of the bazaar, in the traditional Islamic city, as a linear form leading from the city gates towards the city centre where the Jamia mosque is located as a prominent feature in the vicinity of the main Maidan. In contrast, the bazaar of Sanandaj was built in a solid rectangular form (92 m long and 70 m wide) outside the city walls, so that its longitudinal axis is in an east-westwards direction without any direct lines to the caravan roads and even the principal road of the city. More importantly, neither is it in the vicinity of the Jamia mosque, nor does the continuation of Arg square touch it. Such spatial features are a rare feature in Iranian cities and even in the Islamic context. For example, Morris (1994: 390) defines three basic configurations for the Suq/Bazaar in the Islamic city; however, none of them match the form of bazaar in Sanandaj. For Sultanzadeh (2001), Iranian bazaars mainly displayed the two forms of linear and centralised. Even the centralised bazaars, which was further divided into two types, those bazaars that have their Dukans/shops open to the Maidan and those that contained porticos before the shops, do not match the structure of bazaar in Sanandaj, as its shops were arranged face to face along the pedestrian walkway between them (Fig. 3.11).

Bazaar and Mosques

Although the bazaar was detached from the Jamia mosque on grounds of political considerations, there were other mosques close to each gate of the bazaar (Fig. 4.1). This can be recognised as one of the main principles of the Islamic city, as the site of the bazaar/Suq was usually closely connected with the mosques in accessible situations to allow Muslims to perform their daily prayers.

> O you who believe! When the call is made for prayer on Friday, then hasten to the remembrance of Allah and leave off trading; that is better for you, if you know. (Qur'an 49:10 quoted in Memarian & Brown, 2003: 191)

The mosques close to the bazaar act as tranquil islands against the noisy barter of the "tunnel-like bazaar", to withdraw people from the streams of the public life to contemplate for a moment (Bianca, 2000; Browne, 1976). Considering the function of the bazaar, as wholesale centre for widespread areas, these mosques also served as the places of rest and shelter for travellers (Alten, 1958).

Bazaar and Hammam (Public Bath House)

Alongside mosques, the Hammam is another public place situated close to the site of the bazaar. It is a well-known element of the Islamic city, where ritual cleansing prior to the mosque is performed according to Islamic principles (Conner, 1987; Najimi, 1988). It is also "an essential part of the social life in urban Islam" (Sarmento & Kazemi, 2014: 138). This is a complementary urban element corresponding to the function of the bazaar and mosques. The site of the Sanandaj bazaar contained four Hammams close to the mosques (Fig. 3.11). In comparison with other Iranian cities, the number of Hammams associated with the bazaar is a striking feature peculiar to Sanandaj. For example, in the whole of the Isfahan bazaar, along a distance of over a mile, there are only two Hammams, one close to the Maidan-e Naqshi Jahan, and the other in the vicinity of the Masjid-e Jamia (Ardalan & Bakhtiar, 1973). Or, as

Kheirabadi (1991: 54) explains, the city of Sabzivar (in the north-east of Iran near the city of Mashhad) also only contained two Hammams in the whole course of its linear bazaar. More importantly, due to the linear form of the bazaar in other Iranian cities and, thus, its function as the backbone of the city in structuring the neighbourhoods, the mosques and the Hammams could also meet the needs of the neighbourhoods. In contrast, these elements were specific to the function of bazaar in Sanandaj due to its centralised form.

Bazaar and Caravansary

The next element adjacent to the site of bazaar was the caravansary, a public building used for sheltering caravans and other travellers. It was a common feature of the traditional Iranian bazaar, especially in capital cities or cities with the potential of trade activities.

As explained, each neighbourhood was supplied by a Bazaarche/small market, which made the neighbourhoods into self-contained structures. Maintaining such structure led the main bazaar to function as a whole-sale market for those Bazarchaha and the other bazaars in the region (Fig. 3.11). Hence, three caravansaries (Fig. 3.11) were attached to the main bazaar to facilitate the export and import trade activities. More

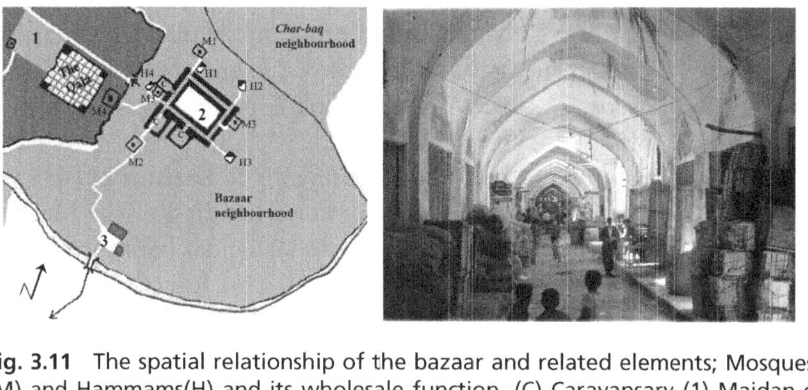

Fig. 3.11 The spatial relationship of the bazaar and related elements; Mosques (M) and Hammams(H) and its wholesale function. (C) Caravansary (1) Maidan-e Arg, (2) Maidan-e Alafha (*Source* Map by author, Autumn 2005 and photo from the collection of Mr. Gazarani, date unknown)

importantly, they were constructed on the south and west sides of the bazaar, away from the course of the principal road. This confirms the claim that the bazaar was not supplied through the principal road. Furthermore, this also means that most of the caravans were directed across two bridges over the Dara-bayan River to the caravansaries; one alongside the eastward course of the city walls, and the second one through the square in front of the church into the criss-crossed lines of the streets.

In contrast with the structural order of the bazaar resulting from its rectangular form, the associated caravansaries did not strengthen this order. This can be ascribed to the different construction times of each of the caravansaries. The architectural structure of each caravansary was based on a courtyard (open space) flanked by chambers and one or two gates into the bazaar and surrounding areas. They functioned as feeder places for the bazaar for the purpose of the unloading and loading of goods.

Physical features of the Bazaar and Maidan-e Alaf-khan

The origin of the rectangular form of the bazaar and its adaptation to the compact form of the settlements and their maze pattern of streets can be traced back to the new concepts of city design in the period of the Safavid dynasty. As Habibi (1996: 95) indicates, in that period (Shah Abbas, 1587–1629), "two patterns of organic and rational were [juxtaposed] in order to project a new concept in design principles of the city". This was recognised as a concept of binarism (duality) which enhanced the richness of Iranian urban architecture (Oliver, 1983). The physical manifestation of this idea can be traced to the spatial crystallisation of the Maidan-e Naqsh-e Jahan in Isfahan and Maidan-e Ganj-Ali-khan in Karman. It can be claimed, therefore, that the structure of the bazaar and its Maidan was derived from these two Mayadin. It is more comparable with the Maidan-e Ganj-Ali-Khan because of its size, the configuration of the bazaar and Maidan, and its location in relation to the Maidan-e Arg (Fig. 3.12). The main difference relates to the connection of the bazaar and the Maidan-e Arg. Such direct connection was not formed

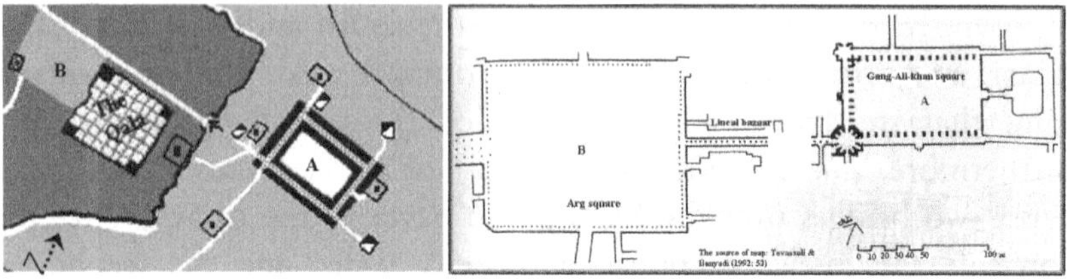

Fig. 3.12 A comparison of the Alaf-khan (A) and Arg squares (B) in Sanandaj with the Gang-Ali-khan (A) and Arg squares in Kerman (B) (*Source* Left drawing by author, Autumn 2005 and right drawing from Tavassoli and Bonyadi [1991: 52])

in Sanandaj due to the barriers imposed by the inner-city wall and the topographical conditions.

The rectangular form of the bazaar gave a rigid structure to the shops (*Dukans*) on four sides, with a face-to-face configuration along the pedestrian walkway in-between (Fig. 4.1). Although the total physical order of the shops, especially the vaulted brickwork structure of the ceilings, reminds us of an imitation of the patterns from the concept of the bazaar in other Iranian cities in the hot-arid zone, it was adapted to the climatic conditions of the region. The point of adaptation is clear from the roof of the bazaar. Contrary to the inside image of the vaulted brickwork ceiling, its external image presents itself as small domes due to the thickness of the roof so as to insulate the inside from the harsh winter weather.

Based on the Russian map (dated 1850), the bazaar and Maidan had four gates in the four sides of its rectangular form. These gates were not aligned exactly with the cardinal points of the compass. As was said before, this was due to the general topographical orientation of the city which gave the main elements of the city a few degrees of deviation from the cardinal points of the compass and thus, a strong correlation with the direction of the Jamia mosque (the holy direction of Qibla). From this point, the longitudinal axis of the Maidan was oriented to the south-east towards the hill of *Toos-nauzar* and thus its latitudinal axis was adapted to the Qibla direction. This can be seen in the orientation of the Jamia mosque (Fig. 3.4 and 3.14). More importantly, since the bazaar is older than the Jamia mosque, one might claim that this order was maintained not just by the effect of topographical features but rather in respect of the

Qibla direction. This claim needs further careful investigation through historical records and archaeological studies, which is beyond the scope of this book.

Function of the Bazaar

As was said before, the bazaar acted as a centre for wholesale activities. In the viewpoint of Ayazi (1992), the products of the region, including gall-nuts, gum-tragacanth, some medicinal herbs and agricultural products (mostly from the west and south), were transferred to other Iranian cities and some to the cities of Sulaimani, Mosul and Baghdad in Iraq by the muleteers of *Qatar-chyan*. On returning, they also supplied the orders of merchants. In the words of Vasilyeva (2000), the most demanded and popular product from Kurdistan was butter. As Tazkirat al-muluk (quoted in Vasilyeva, 2000: 10) indicates, "the Ardalan rulers sent the butter to the Safavid court as a compulsory annual gift – the best of what was grown and produced in Kurdistan". In line with the important role of the bazaar in the life of the city, one can say that "its functioning actually incorporated at least half of the city population: two of the four town sectors were known as Mahalla-ye Bazaar (Aqa-Zaman) and Mahalla-ye Qatar-chyan, the quarter of the transport people, were directly involved in the commercial activities" (Vasilyeva, 2000: 10).

With the foreign trade, similar to other Iranian cities, there was a profession called Sarrafs—money changers. They usually undertook the most complicated banking operations in order to ease the affairs of trade. Convincing evidence of this can be found in the words of Rich (1836: 209), who visited the city in 1820. He mentions the name of a "banker" whose daughter became a wife of Amanullah-khan the great. Ayazi (1992: 493, 494) also explains the role of one Kalimi (Jewish person) as a banker in the Caravansary of Wali. Foreign trade was usually conducted in the caravansary chambers.

According to Ayazi (1992), there were four bazaar lines, known as Char Guzar-e Amanullah-Khani, occupied by different professions:

> The south line, which was named Guzar-e Zahabi-Duzha[5] was occupied by the shoemakers. The west line was allocated to the haberdashery and grocery traders and was known under the name of Guzar-e Kharazi. The north line was divided between the two handcrafts of the saddlers (Guzar-e Sarajha) and other shoemakers, and was known under the name of Guzar-e Uruci-Duzha[6]. Finally, the east line was again occupied by two crafts, that of the goldsmiths (Guzar-e Zargarha) and blacksmiths and was known as Guzar-e Chelangrha. (Ayazi, 1992: 522–535)

It is thus possible to detect the spatial hierarchical order of crafts and trades in the whole structure of the bazaar and to discern that people of the same profession usually occupied the same particular part of the bazaar.

Such hierarchical order can be seen at the west side of the bazaar, where haberdashery and grocery traders linked to the mosques through two Bazaarche in the west and north. These Bazarchaha contained small grocery shops linked to the mosques (Fig. 3.13). This means that the noisier crafts were kept further away from the mosques and more importantly from the seat of power in the west. Although this to some extent conforms to other Iranian cities, here there were no shops selling candles and "*Muhrs* (small tablets made of clay from sacred cities") (Kheirabadi, 1991: 65) close to the mosques, as such shops are specific to the Shiite tradition. The majority of the city's inhabitants adhered to the Shafi'i school of law, one of the four established schools of Sunni Islamic jurisprudence, as they are now. In contrast to Shiite Islam, the Sunni Islamic tradition has no strong connection with Imams and their relatives as continuations of the line of the Prophet Mohammad. Thus, their *Mazars*/shrines, where the people usually light candles to represent their *Nazr*/vow, were not evident in Sanandaj. Therefore, such types of shops and even the Iranian urban element of Hussayniyyah are absent at the site of the bazaar.

Concerning the function of Maidan-e Alaf-khan, apart from the work of Ayazi (1992), there are no more information in the historical books.

---

[5] As Ayazi (1992: 523) explains, "Zahabi-Duzha made special shoes of thick leather that were usable for inhabitants of the rural areas, especially nomadic people and herdsmen".

[6] This was another group of craftsmen that made shoes that were suitable for everybody.

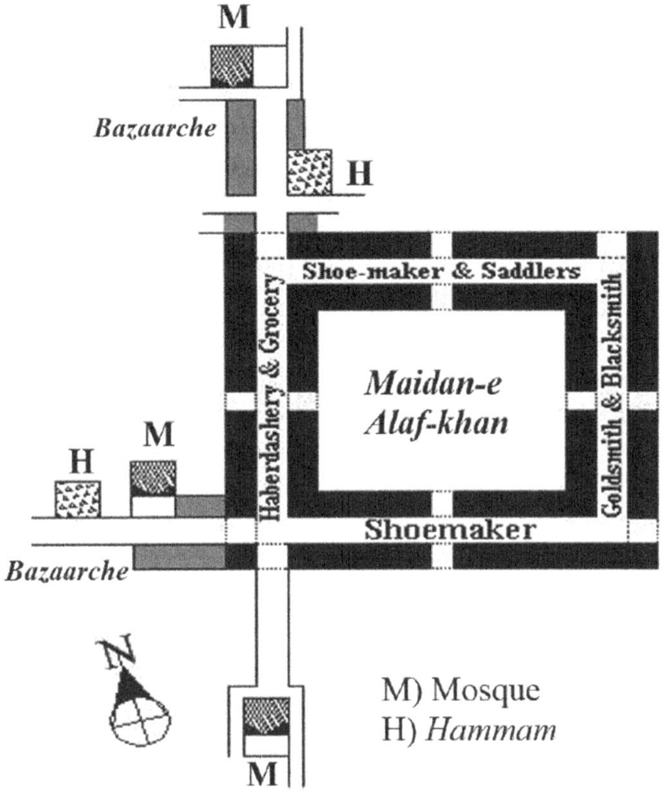

Fig. 3.13 Spatial distribution of the crafts within the four lines of the bazaar (*Source* Author, Autumn 2005)

The explanations he offers also do not comprise an exact function for the Maidan, especially before the construction of the main crossroads in 1938–1939. Afterwards, the crossroads, bazaar and Maidan were separated into two parts along the longitudinal axis. Consequently, nothing is left of the Maidan's open space, as it was occupied by different new shops on both sides of a new urban element named *Khyaban*-e Cyrus (Cyrus Street). This generally blurred the boundary and exact physical sense of the Maidan. Regardless of this, it is possible to recognise the structure and function of the Maidan through comparing the results of direct observation of the site with the explanations of Ayazi (1992) and information from the Russian map of 1850. A further aid would be a cross-referencing to the function and the spatial structure of similar Maidans in other cities, especially Maidan-e Naqsh-e Jahan in Isfahan and Maidan-e Ganj-Ali-khan in Kerman.

From the features of the facades of the existing shops, it can be recognised that the perimeter of the Maidan was surrounded by shops backing onto the shops of the main bazaar. This is similar to some extent to the shops facing Maidan-e Naqsh-e Jahan. In the words of Ayazi (1992), the north side of the Maidan was occupied by some craftsmen, namely coppersmiths and blacksmiths, the south part allocated to the wholesale trade of fruits and vegetables, and the craft of dyers. As the Maidan was named Alaf-khan, its open space was occupied by the Alafs, who sold second-hand clothes set on a Kursi (a square table covered with quilts and blankets that can warm the legs and body by placing a brazier under it at winter). From this and from the existing sources,[7] it can be thought that a weekly bazaar was also held at this Maidan, especially on Fridays.

From the above points concerning the bazaar and Maidan, the bazaar complex functioned mostly to supply the monthly and yearly needs of the city's inhabitants and the people of the region. Thus, the Bazaarche at the centre of each quarter and the small bazaars in other cities and villages acted to meet daily and weekly needs. Kostof (1992: 29–37) describes this way of "weaving the local and long-distance activities into the urban form [as] a distinct feature of the business in the Islamic cities".

As aforementioned, at least half of the city's population was engaged in the function of the bazaar. This was a relatively high proportion, which points to the great role of the bazaar in the socio-economic and even political life of the city. For its political dimension, one may look at the War of the Caravansary, which took place in 1855. As most of the local historians indicate, the craftsmen together with the people of the bazaar played a leading role in that war. In the words of Vaqaeea-Negar Kurdistani (quoted in Vasilyeva, 2000: 9) "one of the heads of crafts, who took an active participation in that war, was heavily punished". In addition, this also confirms the well-defined alliance between the Bazaari (people who work at bazaar) and the Ulama (the learned religious elites) because

[7] In its existing structure, the peddlers usually gathered on both sides of Cyrus Street on Fridays.

the social unrest was initiated by the official protest of the Ulama against the decision of the ruler[8] and followed by the Bazaari. This was recognised by scholars as an important role fulfilled by the bazaar in Iranian cities. For example, Kheirabadi (1991: 60) points out the main role of the bazaar in the "constitutional movement of the early twentieth century and during the revolution of 1979" when the people of the bazaar closed their shops to show opposition to the Shah's regime.

Generally speaking, from the above review and the location of the Jamia mosque in the quarter of the ruling class (located away from the city of the commoners and the main bazaar), it can be claimed that, "unlike the popular belief postulated by the model of the Islamic city, it is not the Jamia mosque but the bazaar that formed the centre and actual focus of the city" (Kheirabadi, 1991: 65). More importantly, due to its role as an important place of "social gathering" during the feast days of the Islamic world and Iran (Vasilyeva, 2000: 10), it was a great place of reference in unifying the disparate parts of the city together (Ardalan, 1980b). It furthermore served to integrate the social life of people from the quarters with different social backgrounds (Muslim and non-Muslim) and even the countryside. As Vasilyeva (2000: 10) indicates, the Sanandaj bazaar serves as the main "foothold" of the city.

Maidan-e Kilisa (Church)

The third Maidan was in the neighbourhood of Aqa-Zaman. As explained, this was an open space in front of the church surrounded by the public buildings of the Hammam and *Bazaarche*. It had the church as structural focal point. Regardless of the Bazaarche's function, the existence of this Maidan to some extent recalls the concept of a square in front of the church in medieval towns of European countries. Its importance can mainly be attributed to the dominant role of the Christians in the trade activities of the city. It was located close to the southern road, which crossed the Dara-bayan River in the direction of Kermanshah (Fig. 3.12). The continuation of this road was linked to the site of the

[8] For more information see Mardukh (2000: 404, 405).

bazaar, especially to the caravansaries, by the tortuous streets (Figs. 3.8 and 3.11).

Due to the lack of recorded information and the fact that the existing built form does not yield any further information either, it is not possible to visualise the form of this Maidan exactly. It certainly did not have a regular form similar to the spatial structure of the Maidan-e Alaf-khan, because it was mainly determined by the irregular form of adjacent buildings and thus was formed according to the organic pattern of the neighbourhood.

Religious Centres

> Among the main features of each settlement, city or village, the place of worship has a particular location predominant over other elements. Thus, it is located in the centre of each settlement. (Peernia, 1995: 37)

The next set of places of public life of the city are the religious centres of the mosques, where communal worship, prayer, teaching and other socio-cultural activities are performed. This means that the mosques are not only meeting places for prayer because in the words of the Prophet Mohammad, "the earth has been created for me as a mosque and a place of purity, and whatever man from my Ummah/community finds himself in need of praying, let him pray anywhere" (Hossain, 2015: 1444; Mazumdar & Mazumdar, 2004). They acted as umbrellas where people of common belief could interact for their contemplative activities. Among them, the main one, which possessed a particular character and symbolic importance, was known as the Jamia mosque, where all adult Muslims must gather to address the socio-political issues of their community and pray with each other behind an Imam in a unified direction (*Qibla*) and at a particular time on Friday. This was expressed by Rumi (the greatest classical Persian poet of the thirteenth century) as follows:

> Cathedral mosques were erected so that the whole city might be assembled there; the Kaaba was instituted in order that the greater part of

> mankind might gather there out of all cities and climes. (Jalal al-Din Rumi, quoted in Madanipour, 2003: 210)

For Pope (1967 quoted in Kheirabadi, 1991: 64) "it is veritably the home of all people in serving many religious, political, social and educational purposes", intended to strengthen spiritual and material bonds among the community, as well as fostering brotherhood among all men from different walks of life and standing. In the model of the Islamic city, it was recognised as the pivotal place of social life of the city and, thus, a fundamental part of city planning in all Muslim countries from Spain to China (Ardalan, 1980a: 18). The trace of this point can be found in the historical background of the city in 1720, when the Ottoman Empire captured the city and left their imprint on the city solely by changing the urban element of the Jamia mosque.

Religious belief was discussed by many scholars as "an inflectional factor affecting all aspects of citizens' lives" in the Islamic world (Kheirabadi, 1991: 62). This is mainly so because "Islam is at the same time a religious tradition, a civilisation, and a total way of life because it prescribes the pattern of order for society" (Morris, 1994: 372). Hence, due to the socio-political background of the Islamic religion, its features formed some of the most significant spatial structures in the cities of the Islamic world. In the Iranian context, the Shiite belief, in line with the descendants of Imam Ali (the prophet's cousin), contextualised the religious buildings of Hussayniya ("building for religious mourning") and the shrines of Imamzadehs (the descendants of Imams). The importance of these buildings was such that they often surpassed that of the city centre and even that of the Friday mosque (Kheirabadi, 1991: 69). However, as the inhabitants of Sanandaj were mostly Sunni Muslims, the city did not contain any independent Hussainiya building.

After this general background concerning the importance of mosques in Islamic cities, the next entry focuses on the main religious urban element of the city namely the Jamia mosque. As explained, the other religious elements are examined together with discussions concerning the neighbourhood of *Qatar-chyan*.

Jamia/Friday Mosque

> In order to understand its role in the city, it is important to understand the mosque's position in the overall urban fabric (Khidir, 1998: 46).

Sanandaj, similar to other Islamic cities, possessed the urban element of the Jamia mosque. As explained in the historical background, the location of the Jamia mosque was affected by a political conflict which took place between two rival powers: the Safavid and the Ottoman empires. The previous Jamia mosque, close to the seat of government and the bazaar, was destroyed and the existing one was erected elsewhere, far from the bazaar and the core of the city in the quarter of the ruling class, close to the west gate of the city (Fig. 3.10). This means that the Jamia mosque was detached from the city of the commoners and tended rather to be connected with the ruling class and less with the other people, up to the time of the destruction of the walls. This is somewhat in contrast to the general pattern of the Islamic city and Iranian cities in particular, where the bazaar either was adjacent to the Jamia mosque, or its continuation created a spatial structure for their spatial interaction. This feature of the Islamic cities is made evident by Falamaki (1995: 129): In the cities that were exposed to the invasion of foreigners or to occupation by local governments, the Jamia mosque was a civic unit more so than in other cities. Therefore, it cannot be separated from the bazaar or maintained in further distance from it, because it is the main pillar of the city's religious power.

There is no clear explanation in the local chronicles concerning the life around the Jamia mosque in that period, particularly concerning its location within the city walls. It can be conceived that the rulers were aware of the status of the Jamia mosque as the main core of society. It played a role in the case of uprisings against unpopular decisions or laws made by those with political power, and functioned as a place of higher education for the Ulama in Islamic cities. For greater security, they kept it within the second wall among the ruling classes. People who came from the bazaar and other neighbourhoods thus had to pass through the street in front of the Maidan-e Arg or one of those linked to it, in order to reach the Jamia mosque and participate in Friday prayers. Under these circumstances, the bazaar became the main civic core of the city. It was supplied by a number of mosques, thus compensating for its spatial separation (Fig. 3.13).

The Jamia mosque was situated close to the west gate (the gate of Dare) on the street known as *Kuche Deraz*, which led from the east gate (the gate of *Agha-Rahim*/bazaar). The continuation of that *Kuche*, which crossed the topographical barrier and led through the east gate to the site of the bazaar could nevertheless create a strong spatial movement in the structure of the city. It thus linked three main public urban elements of the city (Fig. 3.10). From this point and regardless of the previous discussions, we can point out another idea in respect of the Jamia mosque's detachment from the site of the Qala/citadel and bazaar.

As most of the main features of the city were renovated or created in the period of Amanullah-khan the great (1800–1825), mainly by imitating the great works of Shah Abbas in Isfahan, our claim is that he intended to expand the arena of the bazaar beyond its rigid structure to supply the potential trade activities of the region. For this reason, he moved the religious centre of the Jamia mosque to the west—opposite the site of the bazaar, as the site of the Qala was in-between, to create a corridor of movement (a linear bazaar) with its socio-cultural potential role, which would link those three elements (Fig. 3.10). For him, this would have been an idea for the future prospects of the city when the barriers, walls and topography could be removed or changed. As explained, this mainly concerns the location of the Jamia mosque along the *Kuche Deraz* and the fact that the Ardalan rulers often intended to create a splendid city in the region similar to that of Isfahan.

Physical Characteristics

Jamia mosque location is in a relatively higher elevation where it can be seen from surrounding areas as a main landmark in the city (Fig. 3.14). In the words of Ardalan (1980a: 21), this means an emphasis on the "symbolic value" of the mosque. Similar to other Iranian mosques, it was surrounded by streets. This was recognised by Peernia (1995: 46) as a principle to keep the mosque free from contamination.

Its structure comprises three main parts: the prayer hall (*Shabistan*) in the west, the courtyard in front of the east Eyvan/porch or great porch, flanked on three sides by the chambers of teaching as the third part. While the main orientation of its overall plan is towards the east, the rectangular form of the prayer hall was directed towards the holy Qibla (Fig. 3.14). The prayer hall contained two porches. The great porch towards the east was flanked by two Guldasteh (minarets) (Fig. 3.14). This was recognised as the main orientation of the city towards the sacred hill of Toos-nauzar. The second porch faced towards the south-west. The main gateway was also highlighted by a small porch from the south-east side, linked to the Kuche Deraz, now *Imam* Street. More importantly, it was aligned with the latitudinal axis of the courtyard directed to the court of ablutions. This is an important symbolic notion of mosque architecture, which informs people of their first duty before proceeding towards the spiritual space (the prayer hall) (Fig. 3.14). From this point, the court of ablution was located in the centre of the courtyard where the axis of the gateway and the great porch, which contained the entrance to the prayer hall, intersected at right angles. The great porch thus acts as a symbol communicating the importance of the spiritual space.

The roof of the prayer hall was supported by 24 pillars, with the intervening spaces structured as small domes. The direction of the Qibla was emphasised by its rectangular form, a small niche in the Qibla wall, and the second porch in the exterior facade. In spite of these three small features defining the direction of the Qibla, the overall orientation of the mosque and relating elements were towards the east[9] so that the eastward orientation exceeded that of the Qibla. As explained, the eastward orientation corresponds with the city's general orientation as observed in the city's main spatial features (bazaar, Qala, garden ensemble of Khusrau-awa and Char-baq) which was mainly imposed by topographical features of the terrain.

[9] As the deviation of the *Qibla* from the south of the campus is 18 degrees towards the west, it is not exactly in the eastward direction.

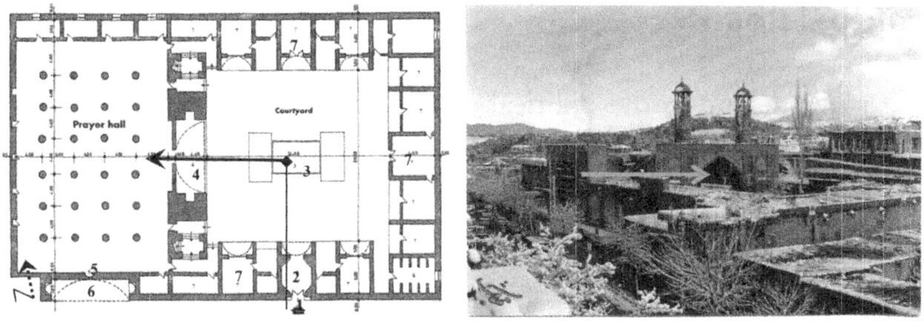

Fig. 3.14 Jamia mosque plan and its image with two minarets. (1) Gateway (2) Hashti (vestibule) (3) court of ablutions (4) Eastern *Eyvan* (5) niche (6) Southern Eyvan (7) school chambers (8) toilets (*Source* Right from Teashk photographer and left from Marjan [1977])

As explained, the courtyard was surrounded by classrooms. There were twelve such chambers to the north, east and south of the courtyard. The number of twelve chambers refers to the twelve stages or levels of knowledge which each Taleba/student had to acquire to achieve the status of *Mallayati* (one who is qualified to teach the holy Quran, give sermons and lead prayers as an Imam in other mosques) in the region).

Neither the existing built form of the city nor the local chronicles provide evidence of any independent urban element of a *Madrasa*. Thus, the Jamia mosque acted as a place of higher education in the region and played an important role in attracting many students from the Iraqi region for higher levels of education. For example, *Mullana-Khaled*, who was recognised as the founder of *Naghsh-Bandya* Tarighat (a spiritual religious sect of Sufism) in Kurdistan, had qualified at that *Madrasa* (Sanandaji, 1996: 65). Because of its status, it was called "*Darull-elm-e* Iran" (Iran's house of knowledge) (Kurdistani, 2002: 33; Sanandaji, 1996: 14).

Most of the local historians explain that the south-west *Eyvan* was due to create the second courtyard for the mosque to contain more classrooms (Mardukh, 2000; Sanandaji, 1996). This can be seen in the small dimension of the entrance door and the south facade of the mosque facing to the public thoroughfare, as it implies more of an interior than an exterior facade (Fig. 3.14).

Principal Road

As explained, due to the political significance of the place and that the role of the *Qala* outweighed its economic importance, the principal road was formed in line with the poles of powers within the quarter of ruling class instead of the bazaar. This created an important character for *Sanandaj* distinguishable from other Iranian cities where the structure of the city is usually bound to the location of the *Jamia* mosque and bazaar.

As the city's backbone, it ran through the entire city leading from the north-east to the south-west. From this point and because it crossed four of the five neighbourhoods, it can be recognised as the most influential morphological component influencing the structure of urban spaces. Its course defines the south-eastern boundary of the *Jawr-awa* neighbourhood as well as the north-west of the *Char-baq* neighbourhood. This means that it functioned as the main thoroughfare, channelling the feeder and collector streets for both neighbourhoods. Along this part of the road, some urban elements evolved, including a *Qahve-khane* (Chaikhana or Coffee shop), mosque and *Amjad*'s house (it was used as a *Hussainiya*). After defining the boundaries of these two neighbourhoods, it crossed the ruling-class quarter through the north gate to the *Arg* square in front of the *Qala* and then to the south gate next to the *Dara-bayan* River. Close to the *Arg* square, *Kuche Deraz* branched off to give access to the *Jamia* mosque. After crossing the entire part of the ruling-class quarter, the road ran along the existing road towards the slope of the Awear Mountains and then to the garden complex of Khusrau-awa (the second seat of power) (Fig. 4.3). This part of the road formed the core of the Qatar-chyan neighbourhood. More details of this part are examined in the following chapter concerning this neighbourhood.

Conclusion

The review above comprised two parts. The first part dealt with the geophysical setting of the city. It presented evidence of the ways in which people related to nature and how the main spatial features of the city were shaped because of the natural features of the landscape. The evidence presented can be conceived as "a tendency in human beings to reshape the features of natural environments within the city's structure. It means that the city takes the principles of its structure from the order of nature and that the structure of natural acts" (Alizadeh et al., 2019: 244) as a necessary first step, namely the foothold for humans to express their presence in the general layout of the city (Lyu, 2019). This means that the spatial layout of the traditional city is more in line with the natural setting compared to cultural values, "an attempt to embrace location and mediate the surroundings" with the manifestation of the content of socio-cultural forces, as a single entity (Leatherbarrow, 1993: 51). Therefore, urban forms of the traditional settlements are highly "place-specific" (Rapoport, 1989: 91), which means that it is first bound up with the natural setting and then determined by social parameters (Alizadeh & Irandoost, 2017; Ozaslan, 1996). In other words, traditional cities' urban form works "with, not against, the forces of nature within its natural foothold" (Kheirabadi, 1991: 39). Therefore, this interaction is a fundamental aspect of "existential space which generally indicates in direct and indirect ways how people have related to the nature and what kind of existence they have created there in order to embrace the outside world, and bring the world close" (Davis, 1997: 140).

The second part dealt with the morphological components of the city. The basic structure of the city was based upon four components (neighbourhoods), each of which was formed by a group of people regarding their professional backgrounds and social status. Although from the topographical features there were only some slight differences between them, their layouts were completely different. For example, the factor of social status limited the ability of choice for the inhabitants of Jawrawa as they had to follow the order of terrain, especially the direction of gullies, in projecting the lines of streets. In contrast, it induced the inhabitants of the Aqa-zaman/bazaar neighbourhood to form the narrow and circular pattern of streets to protect their wealth. Within this part, it became clear that "the urban form is never innocent of social context" (Kostof, 1992: 8). That is why the unique and peculiar features of the place have to be perceived only from local conditions.

In the perceived structure of the city, the court of the Khan on the hilltop rather affected the urban spaces than it did the bazaar and Jamia mosque. The quarter of the ruling class embraced the fortified walls to secure greater privacy from the city of the commoners. For this reason, the course of the principal road was directed alongside the poles of power instead of along the bazaar. Such seclusion of power and of the ruling class from the city of commoners outlined the structure of satellite neighbourhoods, as each contained the element of *Bazaarche* and no direct line linked them to the main bazaar. We therefore observe "village-like communities within the urban whole" (Vasilyeva, 2000: 7). While the urban element of the bazaar was embraced by the compact form of the neighbourhood with no direct line to the countryside or caravan roads, it was located at the bottom of the Qala hill for more control and monitoring.

Due to the results of this chapter, it is possible to say that the city's founders had perceived its main spatial structure in advance as a whole with the design principles serving the interests of the Khans/ruling class. This means that the concept of the planned city can be applied for Sanandaj in a sense analogous to the notion of Akbar (1988 cited in Morris, 1994: 381):

> It is stressed that the term 'planned' applies only in respect of a political decision to establish a permanent settlement on a given site.

References

Alizadeh, H. (2004). The Kurdish city of Sanandaj. *Journal of Kurdish Studies, 5*, 103–126.

Alizadeh, H., & Irandoost, K. (2017). Investigating the process of traditional design principles formation in the Iranian-Kurdish urban quarters. *URBAN DESIGN International, 22*(3), 197–218.

Alizadeh, H., & Habibi, K. (2008). Structural elements and the built environment. *International Journal of Environmental Research, 2*(2).

Alizadeh, H., Bahrami, B., & Farahani, L. M. (2019). Geophysical setting and urban form of a traditional city: Examination of the Kurdish city of

Sanandaj. *Iran. Proceedings of the Institution of Civil Engineers-Urban Design and Planning, 172*(6), 237–247.

Alten, I. (1958). *Sanandaj, a town plan.* United State Operation Mission to Iran.

Amin, A. (2008). Collective culture and urban public space. *City, 12*(1), 5–24.

Ardalan, N. (1980a). The visual language of symbolic form: A preliminary study of mosque architecture. In J. G. Katz (Ed.), *Architecture as symbol and self-identity*. Aga Khan Award for Architecture.

Ardalan, N. (1980b). Places of public gathering. In L. Safran (Ed.), *Places of public gathering in Islam.* Aga Khan Award for Architecture.

Ardalan, N., & Bakhtiar, L. (1973). *The sense of unity: The sufi tradition in Persian architecture.* Chicago, University of Chicago Press.

Arefian, F. F., & Moeini, S. H. I. (2016). *Urban change in Iran.* Springer.

Ayazi, B. (1992). *Ainah-yi Sanandaj.* Ayazi, Burhan.

Bakhtiar, A. A. (2011). Reminiscences of the Maidan-e Shah. In C. Mitchell (Ed.), *New perspectives on Safavid Iran: Empire and society.* Routledge.

Barth, F. (1953). Principles of social organization in Southern Kurdistan. Universitetets Etnografiske Museum. *Bulletin, 7*, 26.

Benton-Short, L., & Short, J. R. (2013). *Cities and nature.* Routledge.

Bianca, S. (2000). *Urban form in the Arab world: Past and present.* Vdf.

Boardman, (1978). *The worlds of Patrick Geddes: Biologist, town planner, re-educator, peace-warrior.* Routledge.

Browne, K. (1976). Making of the city: A special issue on Isfahan. *Architectural Review, 951*(v.CLIX), 259–291.

Bruinessen, M. V. (1978). *Agha, Shaikh and State: On the social and political organization of Kurdistan.* Rijksuniversiteit.

Conner, (1987). On the bath: Western experience of the Hammam. *Culture, Theory and Critique, 31*(1), 34–42.

Davis, H. (1997). Location and site. In Oliver (Ed.), *Encyclopedia of vernacular architecture of the world* (p 140–143). Cambridge University Press.

Dultshahi, A. (1984). *Guqkrapgiai Qkarbi Iran ya Kuhhai Nashnahkti Evista.* Naghshi Jahan.

English, W. (1966). *City and village in Iran: Settlement and economy in the Kirman Basin.* University of Wisconsin Press.

Falamaki, M. (1987). Sheklgiry Shahrhaie Iran (Iranian cities formation). In M. Y. Kiani (Ed.), *Nazari Ijmali bih Shahr-nishini va Shahrsazi dar Iran* (A general study on urbanisation and urban planning in Iran) (p 1–21). Farhang va Ershad-e Eslami.

Falamaki, M. (1995). *Baz-Zindasazie Banaha va Shahrha-ye Tarikhi* (Renovation of buildings and historical cities). Tehran University Press.

Galletti, M. (2001). Western images of women's role in Kurdish society. In S. Mojab (Ed.), *Women of a non-state nation: The Kurd* (p 209–225). Mazda Publishers Inc.

Gaube, H. (1979). *Iranian cities*. New York University Press.

Gehl, J. (2011). *Life between buildings: Using public space*. Island press.

Grandmaison, C. L. C. (1977). Spatial organisation, tribal groupings and kinship in Ibra. *Journal of Oman Studies, 3*(2), 95–106.

Guest, J. S. (1993). *Survival among the Kurds: A History of the Yezidis*. Routledge.

Habibi, S. M. (1996). *Az Shar Ta Shahr* (Analytical review of the city concept and its physical image in the course of time). University of Tehran.

Holland, C., Clark, A., Katz, J., & Peace, S. (2007). *Social interactions in urban public places*. Policy Press.

Hossain, M. Z. (2015). Water: The most precious resource of our life. *Global Journal of Advanced Research, 2*(9), 1–11.

Izady, M. R. (1992). *A concise handbook: The Kurds*. Crane Russak (Taylor & Francis).

Jiven, G., & Larkham, J. (2003). Sense of place, authenticity and character: A commentary. *Journal of Urban Design, 8*(1), 67–81.

Khidir, O. M. O. (1998). *Bridging between the contemporary and the vernacular architecture* (85) (Electronical Dissertation Architecture). Virginia Polytechnic Institute and State University.

Kheirabadi, M. (1991). *Iranian cities: Formation and development*. University of Texas Press.

Kinneir, S. J. M. (1813). [Republished 1973] *A Geographical memoir of the Pesian Empire: Journey through Asia Minor, Armenia, and Koordistan in the years 1813 and 1814*. Arno Press.

Kite, S. (2002). The Poetics of Oman's Traditional Architecture: Towards an aesthetic interpretation. *Journal of Oman Studies, 12*, 133–155.

Kite, S. (2008). Mountain and paradise: Wall and court, space and transition, in Iran and Oman. In H. Alizadeh (Ed.), *International Conference on Zagros Traditional Settlements* (pp. 25–34). Sanandaj, University of Kurdistan.

Kostof, S. (1991). *The city shaped: Urban patterns and meanings through history*. Thames & Hudson.

Kostof, S. (1992). *The city assembled: The elements of urban form through history*. Thames and Hudson.

Kurdistani, A. V. (2002). *Hadigha-E-Nasriah dar Guqkrafia VA Tarikh-e Kurdistan* (History and geography of Kurdistan) (Persian). Tavakouli.

Lambton, A. S. (1991). *Landlord and peasant in Persia: A study of land tenure and land revenue administration.* Martin Press.

Leatherbarrow, D. (1993). *The roots of architectural invention.* Cambridge University Press.

Levine, L. D. (1974). Geographical studies in the Neo-Assyrian Zagros - I I. *Persian Studies, 12*, 19–123.

Lyu, F. (2019). Architecture as spatial storytelling: Mediating human knowledge of the world, humans and architecture. *Frontiers of Architectural Research, 8*(3), 275–283.

Madanipour, A. (1998). *Tehran: The making of a metropolis.* John Wiley.

Madanipour, A. R. (2003). *Public and private spaces of the city.* Routledge.

Madanipour, A. (2013). Public spaces of European cities. *NA, 18*(1).

McDowall, D. (1996). *The Kurds.* Minority Rights Grou

Mardukh, M. (2000). *Tarikh-e- Mardukh (the history of Kurds and Kurdistan).* Karang.

Marjan. (1977). *Tarh-e Jamia Shar-e Sanandaj* (Sanandaj's master plan). Kurdistan Urban Development Organization.

McDowall, D. (2001). *A modern history of the Kurds/translated by Abrahim unis.* Paniz press.

Mazumdar, S., & Mazumdar, S. (2004). Religion and place attachment: A study of sacred places. *Journal of Environmental Psychology, 24*(3), 385–397.

Mehriar, M., Shamil, F., et al. (1998). *Asnad-e-Tasveri shahrhaei Irani (The old maps of Iranian cities).* Shahid Beheshti University Press.

Memarian, G., & Brown, F. E. (2003). Climate Culture and Religion: Aspects of the traditional courtyard house in Iran. *Architecture and Planning Research, 20*(3), 180–198.

Morris, A. E. J. (1994). *History of urban form: Before the industrial revolutions.* Longman Scientific & Technical and Wiley.

Najimi, A. W. (1988). *Herat, the Islamic city: A study in urban conservation.* Curzon Press LTD.

Nikitine, V. (1983). Kurd and Kurdistan. Translated by M. Qazi, Tehran, Nilufar.

Norberg-Schulz, C. (1971). *Existence, space & architecture.* Studio Vista.

Norberg-Schulz, C. (1980). *Genius Loci: Towards a phenomenology of architecture.* Rizzoli International Publications.

Oldenburg, R. (1989). *The great good place: Cafés, coffee shops, community centers, beauty parlors, general stores, bars, hangouts, and how they get you through the day*. Paragon House Publishers.

Oliver, P. (1983). Binarism in an Islamic city: Isfahan as an example of geometry and duality. In *Islamic Architecture and Urbanism. A. Germen (ed). Dammam*, (pp. 121–146). King Faisal University.

Ozaslan, N. (1996). *Historic urban fabric: Source of inspiration for contemporary city form* (373) (Dept. of Architecture. Unpublished Ph.D). University of York.

Pars, B. (2004). *Tarhe Samandah-ye Site-e Firdaus-ye Sanandaj* (Revitalisation of the Sanandaj Firdausi's site). Kurdistan Housing and Urban Planning Organization.

Peernia, M. K. (1995). *Ashnaei ba Meamari-e Islami: Sakhtimanhai Darwn-Shahri va Burwn-Shahri* (Recognising the Islamic architecture). University of Elmuo-Sanat.

Pourjafar, M., Amini, M., Varzaneh, E. H., & Mahdavinejad, M. (2014). Role of bazaars as a unifying factor in traditional cities of Iran: The Isfahan bazaar. *Frontiers of Architectural Research, 3*(1), 10–19.

Rapoport, A. (1989). On the attributes of tradition. In J.- Bourdier & N. Alsayyad (Eds.), *Dwellings, settlements and tradition: Cross-cultural perspectives* (p 77–105). University Press of America .

Rich, C. J. (1836). *Narrative of a residence in Koordistan, and on the site of ancient Nineveh: With journal of a voyage down the Tigris to Bagdad and an account of a visit to Shirauz and Persepolis*. J. Duncan.

Ross, N., & Mohammadpur, A. (2018). Imagined or real: The intersection of tribalism and nationalism in the Kurdish Regional Government (KRG). *British Journal of Middle Eastern Studies, 45*(2), 194–211.

Sanandaji, M. S. (1996). *Tuhfa-ye Nasiriye dar Tarikh va Jughrafiya-yi Kurdistan* (Kurdistan history and geography). Sanandaj, Amir-Kabir.

Sarmento, J., & Kazemi, Z. (2014). Hammams and the contemporary city: The case of Isfahan, Iran. *International Journal of Heritage Studies, 20*(2), 138–156.

Sharafkandi, A. (2002). *Mejwi Ardalan* (The history of Ardalan). Taze Nghah press.

Simmel, G. (2007). The metaphysics of death. *Theory, Culture & Society, 24*(7–8), 72–77.

Spirn, A. W. (1988). The poetics of city and nature: Towards a new aesthetic for urban design. *Landscape Journal, 7*(2), 108–126.

Sultanzadeh, H. (1983). *Ravand-e Sheklgir-ye Shar va Marakiz Mazhabi dar Iran* (The process of City and Religious Centres Formation in Iran). Agah Press.
Sultanzadeh, H. (1986). *Moghadamai Bar Tariekhi Shahr va Shahrnishini dar Iran* (An introduction into the history of the city and urbanisation in Iran). Amir Kabir Press.
Sultanzadeh, H. (2001). *Iranian bazaars*. Cultural Research Bureau.
Tavassoli, M., & Bonyadi, N. (1991). *Urban Space Design1*. Urban Planning and Architecture Research Centre of Iran.
Tavassoli, M., & Bonyadi, N. (1992). *Urban space design 2*. Urban Planning and Architectural Research Centre of Iran.
Vasilyeva, E. (2000). *The Kurdish city of Sanandaj: Ardelan's Capital*. Paris, conference presentation.
Zylstra, G. D. (2017). Productions of space, productions of power: Studying space, urban design, and social relations. *Journal of Urban History, 43*(3), 562–569.

4

Details of Place Structure

Abstract This chapter examines structural elements of the Qatar-chyan neighbourhood in detail and its design principles. The built form of this neighbourhood has been changed less than others and is consequently more appropriate to understand the unfolding of the socio-spatial principles underlying the constituent elements of the traditional built form. This includes the features of the terrain, the socio-spatial factors that gave birth to the neighbourhood, influenced the development and use of urban space and eventually led the neighbourhood to acquire a character distinguishable from other neighbourhoods of the city. This analysis made clear that the main structural elements of the neighbourhood have been bound to three key factors, namely the course of the principal road, the *Wakil* family and the existence of the *Dara-bayan* River.

Keywords Socio-spatial principles · Structural elements · Qatar-chyan neighbourhood · Principal road

H. Alizadeh, *Building a traditional Kurdish City*,
https://doi.org/10.1007/978-981-16-3634-9_4

Introduction

After exploring the city's morphological components, its main public spaces, and the forces and principles underlying their organisation, this chapter focuses on one of the neighbourhoods to make clear the details of place structure and its design principles. In this part, our discussion will be based on existing information, the views of interviewees, historical records, and a wider urban analysis of the *Qatar-chyan* neighbourhood, the built form of which had changed less than that of other neighbourhoods. It consequently is more suitable for our purpose of deriving the details of place structure. It examines the structural elements of the neighbourhood, including the features of the terrain, the socio-spatial factors that gave birth to the neighbourhood, influenced the development and use of urban space, and eventually led the neighbourhood to acquire a character distinguishable from other neighbourhoods.

The Qatar-Chyan Neighbourhood

The word *Qatar-chyan*, as it is used at present together with the word *Janat* (heaven), refers to the area of the old town lying to the south, on the southern side of the *Dara-bayan* River, facing the quarter of the ruling class. It was separated by this river from the traditional part of the city, including the ruling-class quarter and other neighbourhoods (Fig. 3.5). The course of the *Dara-bayan* River created two spatial halves in the structure of the traditional city, which is evident in the old maps, including the Russian map (1850), the Alton Map (1958) and the aerial photograph (1957). Although the river acted as a separation element, it can be recognised as a longitudinal axis that invited man to gather and live close to it. The river also gave the neighbourhood an important potential to acquire a particular character distinguishable from that of other neighbourhoods.

Etymology of the Word Qatar-Chyan

There are different local views on the literal meaning of the word *Qatar-chyan*. In the view of Mardukh (2000), the name *Qatar-chyan* derives from two words *Qater* (mule) and *Chyan* (keepers). Hence, it refers to the main profession of the residents who kept mules there for transportation. This was criticised by Maarufi.[1] He referred to the current spelling of the word *Qatar-chyan* (*Qatar* = train and *Chyan* = set). This meaning still corresponds with the residents' main profession but its literal meaning is different as it implies the arrangement of mules in a line for transportation. More importantly, in one of the historical records concerning property transactions dated 1894, the name of *Qater-chyan* was used in a sense similar to the viewpoint of Mardukh.

Apart from transporting local products to other cities and vice versa, they also provided the city with building materials and other needs from the *Awear Mountains*. This can be seen in the following words of Sheikh Abrahim[2]:

> The train of mules headed towards the Awear Mountains during spring, mostly from the end of winter, to bring snow. The snow was usually stored in special places underground in the neighbourhood, to be used during summer. Because of that, the name Ban Yakhachal (the place of the ice house) was usually used for such a place in the neighbourhood close to the mansion of Wakils' family.

The ice house was specific to the *Qatar-chyan* neighbourhood because of its location close to the slope of the Awear Mountains and the snow pits at its summit.

[1] Interviewed, 15.08.2004.

[2] As one of the interviewees (17.09.2004), he was over 80 years old at the time. As a famous traditional orthopaedic surgeon, he was familiar with most of the older residents of *Qatar-chyan*.

Inhabitants and Their Livelihood

According to local information, herder communities lived in the Qatar-chyan neighbourhood, where they could feed their cattle on the slopes of the Awear Mountains, going out to the slopes at sunrise and returning home at sunset.

> I remember the time that the *Kucheha-ye Mahalla* (neighbourhood's alleys) were crowded each day at the time of sunset with shepherds. (Sheikh Abrahim[3])

Due to the abundance of water from the *Awear*[4] Mountains, the pastures surrounding the neighbourhood were productive. In comparison with other neighbourhoods, which lacked such opportunity, this alongside the self-reliant status of the neighbourhood is the main reasons why some of the families used to keep cattle, sheep and mules.

However, certain families did not engage in this profession. As *Galadari*,[5] whose family was well known for keeping sheep and mules for trade activities, explains, his house contained two separate parts; one comprised living quarters in the north part of the plot, the second part had two stables for about 20 to 25 mules and eight *Zakhaha*[6]—cave-like structures—for sheep in the south. Due to the lack of recorded information from the maps and writings and to the physical changes in the size of the houses, there are no existing signs to show such structure in the houses of the neighbourhood, especially the element of *Zakha* in the spatial structure of the houses.

---

[3] Interviewed 17.09.2004.

[4] The word *Awear* means "the source of water", the place where water originates from.

[5] He is one the interviewees (11.08.2004) and was about 67 years old. His family is known in the neighbourhood for keeping mules and sheep to such an extent that the western part of the *Qatar-chyan* is known as the *Mahalla-ye Gala-darha* (who looks after the cattle and mules).

[6] The *Zakhe* is a built structure, to some extent similar to the ice house used on the Iranian plateau. Its main part is usually underground with relatively more volume than the top, which is narrowed towards the ceiling. Subterranean canals create links to other *Zakhe* or to the outside.

More importantly, as Ayazi (1992), Haqshenas (1999) and Rangavary (1999[7]) make clear, most of the city's gardens (*Baqat*) were located on the slopes of the *Awear Mountains* and some along the *Dara-bayan* River. As explained, the existence of those gardens close to the neighbourhood together with an abundance of water caused Qatar-chyan to be regarded as the place of *Janat* (heaven) by the people. Ayazi (1992) furthermore indicates that some people could find another source of income there in being employed in fruit selling. It was also a place of recreation for young and old alike, gathering there and spending their leisure time in the *Qahva-khane*—coffee house/Chaikhana (Teahouse). This point will be discussed further in the next chapter, which deals with the public spaces of the neighbourhood.

Small-scale handicrafts, such as carpet weaving, were a common source of income for inhabitants of the neighbourhood. It was mainly the responsibility of women to weave the carpets and even to spin its primary materials. They either learned it from family members as a skill handed down over generations or from neighbours in a traditional way of learning from a master:

> The carpet weaving craft was the main honour of the *Qatar-chyan*'s girls. It was an important point for them, as an advantage, to get married. In the past, due to the lack of school for the girls, they had to learn such a craft as a complementary source of income for their family and later on for themselves. (Ayazi, 1992)

He also mentions another type of carpet made by these craftspeople, namely the prayer rug (*Souja*), which was a main part of girls' *Jahaz*/dowry. For this reason, it was necessary for them to gain such traditional knowledge, in order to weave the main item of their *Jahaz*. They could thus also contribute to the income of their families. This shows that women in the commoners' families had a major role in the family's livelihood alongside the men, who were mostly involved with the business of transport.

---

[7] He is one of the master builders (craftsmen) in *Gatch*-work/plastering and was above 90 years old in 1999, his narration, mostly his memories, concerning the past life of the city was tape-recorded.

Some of the inhabitants worked as dealers and shopkeepers of the neighbourhood as they even traded with other cities and into the countryside. Due to the self-reliant status of the neighbourhood, they had a direct link with the villages in the south and west, to supply their needs and at the same time to provide for the agricultural needs of the inhabitants and other cities' neighbourhoods.

Apart from the above professions, some people were engaged in governmental and administrative affairs, including the *Wakil*, *Baban* and *Daroghe* families. In this category, but with less official status, we can mention the *Ulama* (learned people), who were mostly involved in education and religious affairs. As was mentioned before and as will be discussed later on, these two groups as notables of the neighbourhood had an important role in the mobilisation of the people and the administrative affairs of the neighbourhood. Based on their status, their lifestyle and values were rather different from those of the commoners and this resulted in different attitudes to the spatial structure of their houses. This will be examined in the following chapters. Considering such status and preferred cultural values, the women of this stratum were helped by others, mostly by servants and people named *Zheer-mala*,[8] and thus, their territory was limited to their personal realm mostly within the private domain of the houses.

Formation and Development

Before moving on to the structuring elements of the neighbourhood, its process of formation is examined through morphological analysis to explain the neighbourhood's evolution over time. This analysis reveals various socio-spatial factors that affected its built form in conjunction with driving forces.

[8] A poor family who lives in the house of a rich or notable family. They usually reside in a part of the house close to the entrance door. The details of this will be expanded in the following chapter.

First Phase: Qula Mosque

As explained earlier, based on the survey by the Kurdistan Cultural Heritage Organisation (KCHO), the date of the *Qula* mosque's construction can be traced back to the period of the *Safavid* dynasty (1502–1736). As the city was founded in 1639, the *Qula* mosque is one of the oldest buildings in the city (Fig. 2.1, left map). This fact also indicates that the settlements around the mosque, on the slopes of the *Dara-bayan* River, are in the main the older parts of the neighbourhood with the possibility of a close link with other settlements on the opposite side of the river outside the city wall.[9] The mosque must thus have been the first public gathering place or the core of the neighbourhood in that period facing the *Darogheh*[10] alley. Its character and structure are largely similar to those of mosques in rural areas, with a simple appearance and no symbol indicating that it is a mosque and distinguishing it from the surrounding houses.

Second Phase: The *Wakil* Family

The second phase of neighbourhood development concerns the relocation of the *Wakil* family[11] in the southern part of the neighbourhood. Based on the sayings of Mardukh (2000), this family was in the deputy position to *Ardalans*' government until 1749. A conflict occurred when a war took place between *Wali Ardalan* and his opponents, the *Baban* family, in 1750. As the *Ardalans* were defeated due to the failures of *Abrahim bag-e Wakil* (the head of *Wakil's* family), this family was deprived of this position. More importantly, this pushed them to leave

[9] As explained in the historical background, before the city's foundation there were a number of settlements on the slopes of the *Qala* hill (citadel) which were called *Senna* and which belonged to the *Zareen-kaoush* (golden shoes) family.

[10] The Darogheh, as an official person, was usually associated with public order and security in traditional Iranian cities.

[11] The *Wakil* family was one of the noble families in the quarter of the ruling class (*Meyan-qala*), but was driven out of that quarter because of continuous opposition to the ruling house (*Wazier* family) and acting on side of the *Ardalans*' adversaries.

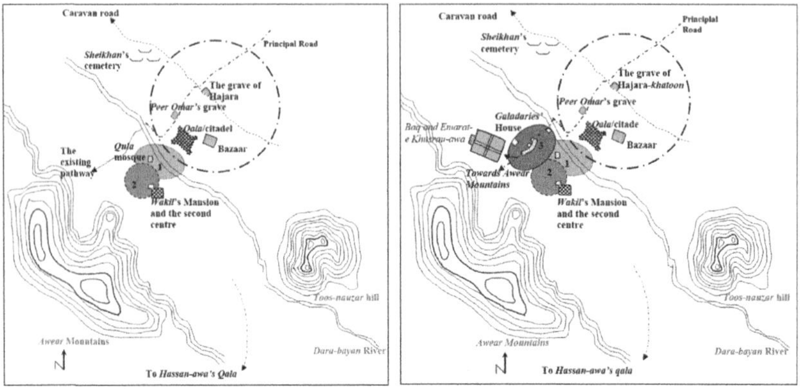

Fig. 4.1 Left, the second and right are the third phases of neighbourhood development in conjunction with *Wakil*'s family and motivated by different socio-political factors (*Source* Author, Autumn 2005)

the city and join *Karim Khan-e zand*.[12] Babani (1998: 53) confirms this to some extent:

> After the reconciliation of *Wali Ardalan* and *Wakil* in 1785, five thousand *Toomans* were given to *Wakil* in order to renovate his house, which was completely destroyed during his exile.

The *Wakils'* relocation in the southern part created the second phase of neighbourhood development (Fig. 4.1). To some extent, the place where they settled defined the southern boundary of the neighbourhood and acted as a pole of gravity to attract more people to settle in their proximity. In addition, their location attracted some other important families, most notably the *Baban* family, who were opposed to the *Ardalans*. Furthermore, their mansion became a reference point for the second centre of the neighbourhood called *Wakil's Bazaarche*. This means that the first centre of the neighbourhood in the older part adjacent to the *Qula* mosque gradually moved over time to this new location

[12] As Mardukh (2000: 351) indicates, due to that event, *Abrahim Bag-e Wakil* encouraged *Karim Khan-e Zand* (the founder of *Zand* dynasty 1757–1794) to capture *Sanandaj*. As explained, he captured the city and set it on fire in 1751.

defined by the mansion of the *Wakils*. More importantly, "this makes clear that the place of gravity or the concept of centre (merging points) was mainly bound to the presence of key figures, the mansions of the Wakil family" (Alizadeh & Habibi, 2008: 157). This is in line with the view of Tjahjono (1989) who says that: "The idea of centre was historically manifested in the ruler who represented the state which in turn symbolized the cosmos. The cosmic power was oriented to, stored in, and radiated from the ruler. The ruler became an ideal to be followed, and his dwelling set a standard to be pursued".

Third Phase: Multi-Factors

The third phase of neighbourhood development can be linked to the main profession of the residents, particularly to the relocation of the *Galadarie* family (cattlemen) in the neighbourhood. As the main profession of this family was cattle herding, they settled in the far western part of the neighbourhood. Over time, they gradually became engaged in transport business linked with the function of the bazaar, as well as daily and seasonal matters of the city. Thus, to meet the needs of the bazaar, they added the profession of muleteers to that of herdsmen. Engaging in that business meant that *Galadarie's* family employed more people to run and boost their business. This led to the relocation of those people in the neighbourhood and consequently to the development of the neighbourhood towards the west.

The above process together with the creation of the *Baq* and *Emarat-e Khusrau-awa* (the mansion and garden complex of *Khusrau-awa*) in the period of *Amanullah-khan* the Great (1800–1825) further pushed the neighbourhood's development towards the west. The garden complex became the second point of gravity for the continuation of the principal road towards the *Awear Mountains* by building a bridge over the *Dara-byan* River. From a spatial perspective (Lawson, 2001), this garden complex can symbolise and communicate the role of its occupants and also the dominant power of its creator throughout the city and the region of Kurdistan. The course of this road towards the garden led some important families, mostly *Ulama* (educated people with religious backgrounds) to settle along the road. These developments provided the basis for the third centre of the neighbourhood, called *Qatar-chyan Bazaarche* (Fig. 4.1), along the road in a linear form.

Considering the above review, the neighbourhood's formation and development took place in three stages regarding different socio-political factors. The first stage refers to the first period of the city's foundation and possibly before it. At that time, the focal point of the settlements was closely linked with the *Qula* mosque. The second stage relates to the relocation of *Wakil's* family in the southern part of the neighbourhood. In the third stage, several socio-political factors affected the neighbourhood's development towards the west, and more importantly, the creation of the major linear *Bazaarche*. These factors include the existing course of a road to the *Awear Mountains*,[13] the economic role of the *Galadarie* family in employing more people in the profession of transportation, the political decision of the ruler in creating a prominent symbol of power across the city along the pathway, and finally the settling of important religious families along that road.

Notably, in the first and second phases of neighbourhood development, the morphology of the centres formed a concentrated structure as the point of reference and convergence was bound to one factor. In the third phase, however, the morphology of the neighbourhood centre changed to a linear form (*Qatar-chyan Bazaarche*) due to the influence of different socio-political factors along the road. Hence, the formation of the linear *Bazaarche* can be compared with the concept of point and line in the work of Ardalan and Bakhtiar (1973: 89). As they argue, "the centre as a single point in space moves in time and creates the line". For Sanandaj, this means that the existing passway to the Awear Mountains was intensified in time by the movement of a centre as a single point in space—the qala—to the slope of these mountains which created another sign of ruler presence—the mansion and garden complex

[13] As the course of this road earlier also was the main pathway towards the *Awear Mountains*, we can assume that it was recognised by the *Ardalan* family as a potential pathway to build their summer house and to define a new point of reference in the structure of the city to emphasise their power. This is in line with the view of Norberg-Schulz (1980: 17) that "where nature indicates a direction, he (man) makes a path".

of *Khusrau-awa*—and attracted other activities along the line and formed *Qatar-chyan Bazaarche* (Fig. 4.2). In this comparison, the point of difference is in the poles of gravity. As they argued, in Iranian cities, those poles were bound to the palace precincts and the city's symbolic heart (*Jamia mosque*) which eventually ended at one of the city gates along which "the vital organs of the city" including the most "religious, political, financial, and social integration of the traditional city" evolved (Ardalan & Bahktiar, 1973: 93). In *Sanandaj*, in contrast, although the first pole of that movement was in relation to the power structure focused on the *Wali's Qala*, the second pole was not bound to the *Jamia* Mosque. It was likewise related to power relationships. For this reason, the vital backbone of the city did not include the structure of the main bazaar, and it just formed alongside the poles of power linked with the political dimension of the city.

As was mentioned before, the detachment of the bazaar from the principal road took place on account of the pre-eminence of the political dimension of the city over other socio-economic factors. Consequently, the city's backbone was directed alongside the poles of power (the *Qala* and the garden complex of *Khusrau-awa*). Along that road the Qatar-chyan linear *Bazaarche* (small bazaar) also evolved. Therefore, it is logical

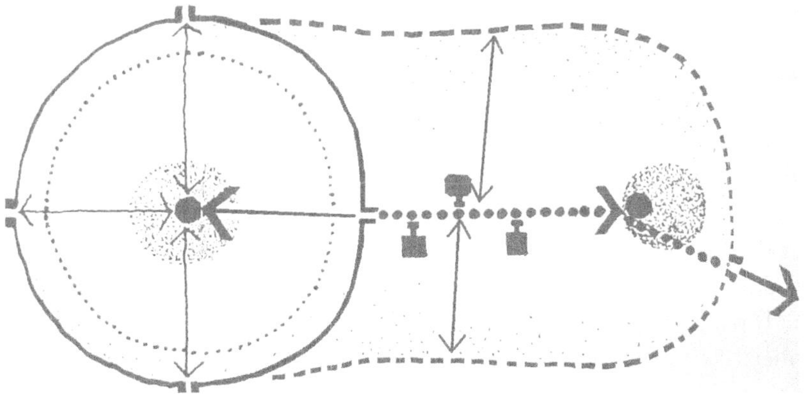

Fig. 4.2 The concepts of point and centre in *Sanandaj which formed the course of principal road* (*Source* Adapted from Ardalan and Bakhtiar [1973])

to infer that the role of power relationships had a major influence on the structure of urban spaces in Sanandaj.

Structural Elements

From the above review, it is possible to infer the structural elements of the neighbourhood regarding the following socio-spatial factors.

First: Principal Route

Based on the morphological analysis of the city, the main structural element of the neighbourhood is a part of the principal road in a north-east-south-westward direction, which crossed the whole neighbourhood and divided it into two separate eastern and western parts, which ended at the summer mansion of the ruler, *Baqand Emarat-e Khusrau-awa* (Fig. 4.1). Along this linear street, the centre of the neighbourhood evolved at the junction points of local streets.

The ruler usually spent his summer time close to the slopes of the *Awear Mountain* in the south where the main greenery and recreational spaces are located. His main way to this site was through the principal road, by crossing the southern gate of the ruling-class quarter, a bridge over *Dara-bayan* River and the *Qatar-chyan* linear *Bazaarche*. It also affected the neighbourhood evolution and even intensified its main centre. More importantly, it was used by Reza-khan when he visited the city for the first time after his coronation in April 1926. Through this road, he first entered the citadel and from there he continued his way through the road towards the south-western part of *Qatar-chyan* to parade his troops and meet people in *Kalaka-jar*[14] (Rangavary, 1999). Madanipour (1998: 226) indicates that "the continuity in [the principal road] of the urban structure signifies the power relations focused on the [ruler], as the highest authority who was far above any other member

[14] At the time, *Kalaka-jar* was arable land, which was later gradually included in the city development, and now it is one of the neighbourhoods that the course of the pathway to the *Awear Mountain* formed, another linear centre named *Kalaka-jar's Bazaarche along Nawroz alley*.

of the administration and played a personal key role in most important affairs of the [region]".

The course of the principal road and its link with the citadel can be compared with the axial structure of Tehran, which started from the northernmost area, where the ruler resided, to the rest of the city in the south, especially to the old part of the city (the *Arg* palace of the nineteenth century). In this comparison, the point of difference lies in the structure of the roads. In Tehran, the course of the road was formed in an axial pattern due to the flatness of the basin that facilitated the connection between the poles of gravity. In *Sanandaj*, this structure was changed to an organic pattern due to the unevenness of the city site (Fig. 4.3). This means that the axial pattern of the urban structure was bound to the topographical conditions of the basin within which the city evolved to a greater extent than in Tehran (Krehl et al., 2016).

Apart from the role of power and centrality in defining the spatial development of the principal road, intermediate agents are also important in this regard (Madanipour, 1998). These agents were called *Ulama*. "As they belonged themselves to different classes", they acted as mediators between the ruler and the ruled, and in linking the parochial groups together (Lapidus, 1967: 114). In spatial terms, the role of these

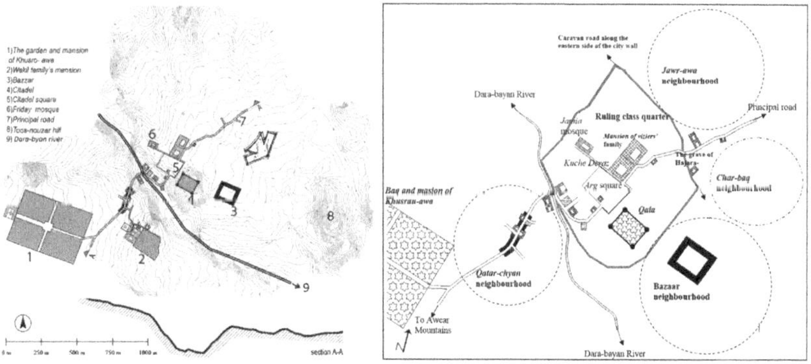

Fig. 4.3 Topographical features of the city site and the course of the principal road and its influential role in the organisation of urban spaces and the city's main urban elements (*Sources* Both by author 2005. The section is from the course of principal route; Topography map sketched by Mr. Shahnazi)

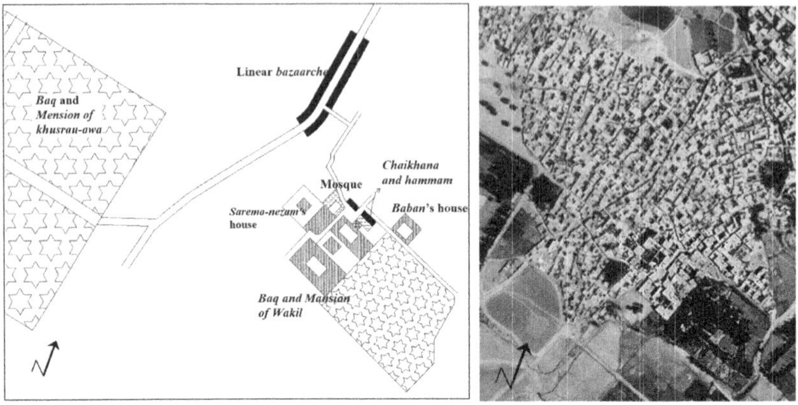

Fig. 4.4 The mansions of the *Wakil* family as the second structuring element of the neighbourhood (*Sources* Sketch by author, Autumn 2005, aerial photograph from INCC [1957])

agents can be traced through the location of the mosque forming the gravity point of the road, which also facilitated the location of the *Qatar-chyan Bazaarche*, and particularly the houses of *Ulama* along the road (Figs. 4.3, 4.4). Their roles were not limited only to the neighbourhood, but extended into the city and even the whole region.[15] The most notable of them were the houses of *Ayatulah Mardukh,* who was the Imam of the *Jamia* mosque and the author of more than 100 books, and of *Mullah Abdul-Azime Mujtahead* and his son *Mullah Sedeeq*, who were the *Mudareseen*/teachers at the *Jamia* mosque.

The location of *Mardukh*'s house along the road is an important element stimulating the *Qatar-chyan Bazaarche* to such an extent that part of the linear *Bazaarche* is sometimes referred to as *Rastae Ayatulah*

[15] To understand how *Ulama* deal with people's affairs, I visited *Mullah Abdul-azime Mujtahed's* son on 22.08.2004. After a short interview, I observed and took notes of his reactions to the people's requests. One of his visitors, who came from the city of Kamyaran to the south of Sanandaj—on the way to Kermanshah, asked Mujtahedi about a problem experienced among heirs. After he gave some clarification, he wrote down his comments as a footnote to a letter on the theme of inheritance. Then, a woman came and explained her case concerning an amount of money borrowed by her husband. As the purpose of the loan was different from what her husband intended to do with it, she asked him for his opinion concerning this change, whether it was right or not.

Mardukh (Series of Shops). As Ayazi (1992: 542) says, "the existence of their houses along the *Qatar-chyan Bazaarche* had created a spiritual atmosphere among the people and *Kasaba*/tradesmen so that on the occasion of prayers, they united in *Tuba-khanoum's* mosque". This atmosphere, he continues, had a major influence on neighbourhood relationships and their social cohesion and solidarity. In addition, many people even from the countryside visited their houses to explore and resolve their socio-political issues. This atmosphere together with the location of the neighbourhood on the slopes of the *Awear Mountain*, as a site of recreational activities, gave the neighbourhood a particular character as the place of *Janat*/heaven.

The linear *Bazaarche*, as a part of the principal road, connected the neighbourhood to the city as a whole, and particularly to the congressional mosque through the southern gate of the inner-shar/town and to the main bazaar along the southern wall. Generally, the connection to the bazaar was bound to the structure of place, both natural and man-made, so that it was formed in defined spaces which facilitated a transition of pedestrian flows from the neighbourhood to the bazaar with minimum intrusion into the territory of other neighbourhoods. This structure reveals that each neighbourhood had a defined boundary with an obvious functional importance for its inhabitants, especially in case of conflict, for instance the War of the neighbourhoods.[16] Apart from the natural structures defining a particular boundary for the neighbourhoods, profession was another factor which had a great influence on the formation and presence of such territorial boundaries.

From the point of view of boundary, the case of the *Qatar-chyan* neighbourhood is somewhat different from that of the other neighbourhoods. Here, the terrain—along the course of the *Dara-byan* River—facilitated a particular boundary or edge for the settlements. The course

---

[16] As mentioned before, the powerful family of the *Wakils* always acted on the side of the *Ardalans*' adversaries. Thus, the political attitude of the people living in these two neighbourhoods (*Qatar-chyan* and *Meyan-qala/Ruling quarter*) might be divergent and in some periods absolutely opposed (Vasilyeva, 2000). Thus, it can be conceived that the war between neighbourhoods, witnessed by Ayazi (1992), might have originated from the attitudes of these two opposing fractions, because most of the clashes happened between the youths of *Qatar-chyan*, as the followers of *Wakil's* family, and the people of *Jawr-awa*, as the followers of the ruling house.

of this river together with the other socio-spatial factors—the existence of *Wakil's* and *Galadarie's* families, *Ulama*, and a part of the principal road which contained an active linear *Bazaarche*—enabled the neighbourhood to be self-reliant. These features created a strong sense of place among the residents, to such an extent that they often differentiated themselves from the other neighbourhoods. For example, one of the interviewees, *Monaghi*,[17] recalls the following quotation from elderly people of the neighbourhood:

> When they wanted to go out of the neighbourhood to the main city, they usually said "let's go to the city".

From this statement, it is clear that their sense of separation from the main city was strong. It means that the inhabitants of *Qatar-chyan* did not consider their neighbourhood as a part of the city but, rather, as an independent settlement detached from the main city. Another convincing witness for this is the war of neighbourhoods within which the neighbourhood of *Qatar-chyan* acted as an independent entity against the other neighbourhoods, mostly the quarter of the ruling class and the *Jawr-awa* neighbourhood (the *Khan*'s servants). In most of the conflicts, *Qatar-chyan* was in opposition to one or more other neighbourhoods. Another highlighting point of such sense of independence is an interesting point mentioned by one of the interviewees, *Ostad Rashid Azhir*[18]:

> The residents of *Qatar-chyan* consulted *Ayatulah Mardukh* about the destruction of the bridge over the river of *Dara-byan* to cut any link with the other parts of the city.

[17] Interviewed 25.08.2004. In the past, he was a craftsman—a cobbler—in the main bazaar.

[18] Interviewed on 06.09.2003. He was a master builder more than 80 years old at the time of the interview. He played a major role in the construction of the city's main the municipality building—the current building opposite the *Jamia* Mosque—which dates back to the period of Reza Khan (1921–1941).

Second: The *Wakil* Family

Considering the findings of the previous section, the next structural element of the neighbourhood is the striking area inhabited by the *Wakil* family in the eastern part of the neighbourhood. As explained, this family was one of the noble families that were driven out from the neighbourhood of ruling classes because of a conflict with *Wali*.[19] Their living mansion in this part was a point of gravity for the second centre of the neighbourhood called *Bazaarchae Wakil*, which contained the main elements of a traditional **Kurdish** neighbourhood including mosque, *Qahve-khane (Chaikhana)*, *Hammam* (bathhouse) and groceries.

Considering the aerial photo of 1957, one especially notes the clear separation of the neighbourhood into two parts by the course of the principal route crossing it like a long crack in the compact form of the settlements. The important point lies in the western part, with a greater area that did not possess a spatial structure similar to *Wakil Bazaarche*. More importantly, the presence of the *Wakil* family influenced the urban grain of nearby houses (Fig. 4.4). In comparison with other houses, theirs show a more coarse-grained structure. Furthermore, such a group of mansions resembles another seat of power which to some degree is set apart from the commoners through a special boundary, including two sides free of buildings to maintain an extensive view to the garden in front and to the green spaces on the slopes of the *Awear Mountains* to the south-west.

From the above perspectives, the presence of the *Wakil* family in the neighbourhood, together with the nearby large houses, can be compared with the seat of power and the ruling houses in the city. This was a real city set alongside the main city separated by the river. The features mentioned again totally support the view that in traditional cities such as *Sanandaj* the role of power relations had a major influence on the structure of place.

The results of this entry show that the place of gravity or the concept of centre (merging points) was mainly bound to the presence of key

[19] As Mardukh (2000) indicates, this family was in charge of the city's agency for a long time and they deserved themselves to possess the deputy position of the Kurdistan government instead of the *Wazieris'* family.

figures, the houses of *Ulama* in the linear *Bazaarche* and the mansions of *Wakil's* family. This to some extent recalls the view of Tjahjono (1989: 233), who says that:

> The idea of centre was historically manifested in the ruler who represented the state which in turn symbolized the cosmos. The cosmic power was oriented to, stored in, and radiated from the ruler. The ruler became an ideal to be followed, and his dwelling set a standard to be pursued.

Third: *Dara-Bayan* River

The third structural element of the neighbourhood is the course of the *Dara-bayan* River, which defined Qatar-chyan's northern boundary. It served as a visible edge to define a clear image of the neighbourhood's territory, especially its northern side facing the quarter of the ruling class. The slope of the river, as an important structure of nature, influenced the morphology of the settlements and contextualised the pattern of terraced houses facing the north with two storeys overlooking a courtyard (*Housh*) and maintaining a broad view of the scenery (Fig. 3.3). An open veranda (*Bar-haiwan*) in the structure of the houses cleared the view towards the distant landscape. This created an outward-looking layout for the houses in accordance with the main feature of the settlements in the region. This corresponds to the view of Bonine (1979), that local topographical features mostly affect the morphology of Iranian cities.

The outward-looking structure of the houses is a spatial feature which distinguishes the settlements of the region from the common structure of the houses in other Islamic and Iranian cities, where an introverted layout shut them off from their surroundings by high and solid walls (Madanipour, 1998: 237; Warren & Fethi, 1982: 44). As explained, this outward-looking feature of the settlements is due to the structure of the place bound to the variation in the surface relief, which created direction and defined spaces in the form of an articulate order.

From the above points, one may think that variation in the surface relief of the terrain was the main motivating factor in the outward-looking structure of the settlements in the city, particularly in *Qatar-chyan*. Although this factor is a basic point that created some similarities among the appearance of settlements in the mountainous areas of Iran with different socio-cultural backgrounds, the socio-cultural circumstances, which made Kurdish society different from surrounding cultures—Arab, Turk and Persian—is also important. This will be further discussed in the following chapter.

Conclusion

This chapter dealt with the structural elements of the neighbourhood through analysing the features of terrain alongside the socio-political factors that gave birth to a built form with a character distinguishable from other neighbourhoods of the city. This analysis made clear that the main structural elements of the neighbourhood have been bound to three key factors, namely the course of the principal road, the *Wakil* family, and the existence of the *Dara-bayan* River. The principal road had a major role in defining the merging points of activities and social life across the city, thus, an active linear *Bazaarche* evolved along and it became the main centre of the neighbourhood. On the other side, the presence of the *Wakil* family created another point of convergence for social activities in the neighbourhood, so that the second centre of the neighbourhood evolved in a concentrated form. The course of the *Dara-bayan* River defined the northern boundary of the neighbourhood and its slope highlighted the architectural element of *Bar-haiwn*—open verandas—in the structure of the houses. It thus contextualised the outward-looking structure in the layout of the houses. This pattern is in sharp contrast to the concept of privacy, which has been recognised as a striking factor in the organisation of Islamic cities (Abu-Lughod, 1983; Alizadeh, 2012; Madanipour, 1998; Memarian & Brown, 2003; Monteguin, 1983; Stewart, 2001). This will be discussed in greater detail in the following chapter regarding the attitudes of Kurdish people to privacy.

References

Abu-Lughod, J. (1983). Contemporary relevance of Islamic urban principles. In A. Germen (Ed.), *Islamic architecture and urbanism* (pp. 64–70). King Faisal University.
Alizadeh, H., & Habibi, K. (2008). Structural elements and the built environment. *International Journal of Environmental Research, 2*(2).
Ardalan, N., & Bakhtiar, L. (1973). *The Sense of unity: The sufi tradition in Persian architecture*. University of Chicago Press.
Ayazi, B. (1992). *Ainah-yi Sanandaj*. Ayazi, Burhan.
Babani, A., (Ed.). (1998). *Saerull-Akrad dar Tarikh va Guqkraphia-E- Kurdistan* (An overview of history and geography of Kurdistan). Awrang Press.
Bonine, M. E. (1979). The morphogenesis of Iranian cities. *Annals of the Association of American Geographers, 69*(2), 208–224.
Haqshenas, M. (1999). *Sharakam Sennah* (Sanandaj; my land). Zhyar.
Krehl, A., Siedentop, S., Taubenböck, H., & Wurm, M. (2016). A comprehensive view on urban spatial structure: Urban density patterns of German city regions. *ISPRS International Journal of Geo-Information, 5*(6), 76.
Lapidus, I. M. I. M. (1967). *Muslim cities in the later middle ages*. Harvard University Press.
Lawson, B. (2001). *The language of space*. Architectural Press.
Madanipour, A. (1998). *Tehran: The making of a metropolis*. Wiley.
Mardukh, M. (2000). *Tarikh-e- Mardukh* (The history of Kurds and Kurdistan). Karang.
Memarian, G., & Brown, F. E. (2003). Climate culture and religion: Aspects of the traditional courtyard house in Iran. *Architecture and Planning Research, 20*(3), 180–198.
Monteguin, F.-A. D. (1983). The essence of urban existence in the world of Islam. In A. Germen (Ed.), *Islamic architecture and urbanism* (pp. 43–63). King Faisal University.
Norberg-Schulz, C. (1980). *Genius Loci: Towards a phenomenology of architecture*. Rizzoli International Publications.
Rangavary. (1999). City's memories/Interviewer: M. Haghshenas, S. Marufi, & S. Munaghi. (Vol 1). Sanandaj.
Stewart, D. J. (2001). Middle east urban studies: Identity and meaning. *Urban Geography, 22*(2), 175–181.

Tjahjono, G. (1989). Centre and duality in the Javanese Dwelling. In J.-P. Bourdier & N. Alsayyad (Eds.), *Dwellings, settlements and tradition: Cross-cultural perspectives* (pp. 213–236). University Press of America.

Vasilyeva, EI. (2000) The Kurdish city of Sanandaj: Ardelan's Capital. Paris, conference presentation.

Warren, J., & Fethi, I. (1982). *Traditional houses in Baghdad*. Coach.

5

Public Domain

Abstract Chapter 6 discusses the domain of public space encompassing the social life of the neighbourhood to find out how the society organised itself in the level of neighbourhood. This includes both social and physical dimensions to approach the design principles of the spaces and also the ideas and beliefs which contextualised them in that society. The way space is subdivided and the relationship between the public and private spheres in general is a mirror of social relations and a main indicator of how a society organises itself. The perceived socio-cultural background and the spatial features show that Kurdish culture is responsive to the concept of Genius Loci, and the dwellings in particular have a strong relationship to the landscape. It was also found that the notion of social life in a Kurdish city is rather different from other Islamic cities due to the socio-cultural and environmental circumstances that have embraced this society and engendered a distinct notion of neighbourhood among its members, and thus, different attitudes to public life in respect to women's presence. Concerning the physical form of the neighbourhood, these attitudes together with the features of the terrain gave rise to a soft boundary between the two realms of social life: the public and the private domains.

H. Alizadeh, *Building a traditional Kurdish City*,
https://doi.org/10.1007/978-981-16-3634-9_5

Keywords Public space · Social factors · Mewan-khana (guest house) · Chaikhana (teahouse) · Visual privacy · Women's public place

Introduction

"As cities of all cultures and the social life of their citizens, at all historical periods, are organized along some form of public–private lines" (Madanipour, 2003: 1), after analysing the structural elements of the neighbourhood, this chapter and Chapter 6 discuss the two domains of public and private spaces encompassing the social life of the neighbourhoods to find out how the society organised itself in the neighbourhood. This will be examined from both social and physical dimensions to approach the design principles that aggregated these realms and also the ideas and beliefs which contextualised them in that society. The way space is subdivided and the relationship between the public and private spheres in general are a mirror of social relations and a main indicator of how a society organises itself.

As most of the literature concerning Islamic cities dealt predominantly with the male's domain, in this chapter, the woman's domain is highlighted and the main points are organised in relation to the role of women in the social life of the selected neighbourhood regarding the following questions: do women relate to the public sphere in ways other than the common view of unsafeness? How do men relate to domestic space? What are gender relations within the household? Do Kurdish public/private spaces differ from those in the rest of Iran? A comparative perspective with Iranian and Islamic contexts is adopted to make sense of the features of public life in this context.

Considering the structure of the book which starts from the whole to the parts, first, the discussion deals with public spaces of the neighbourhood, and then it moves into the private spaces of the houses in the following section as Chapter 6. Between the two realms, within this chapter, the cluster of houses are analysed to discover the manner of aggregation, neighbourhood structure and the transition to the private domain.

Public Domain and Social Factors

The notion of public life and its physical manifestation in the neighbourhood must be viewed from two contexts due to the nature of monograph which deals mainly with Kurdish culture located within the macro-context of Islamic-Iranian cities. One side emerged from the Islamic rules which advocate its members to spend their leisure times concerning the religious affairs of the whole community, the second one originated from the socio-cultural and environmental circumstances which affected this society to adapt themselves with particular secular social activities specific to the Kurdish culture.

Religious Elements

Similar to other Islamic cities, the mosques, as the main public buildings and, thus, as an integral part of the neighbourhood through which the daily public gathering of the adult Muslims are facilitated, maintained and boosted the nature of social life among its inhabitants. The effect of the mosques was enhanced by the presence of the complementary element of *Hammam* (public bathhouse) "a place to provide the sacred ablution required prior to praying" (Kheirabadi, 1991: 82) and the houses of *Ulama*, a place of knowledge to resolve problems and disputes among the inhabitants.

Islamic principles require that each Muslim should pray five times a day. This imperative along with the spiritual value of praying collectively necessitated daily gathering so that it facilitated a possibility for the inhabitants to meet each other daily (Kheirabadi, 2004). In the light of this idea, accessibility was an important principle so that the mosques were mostly located at the merging points or on the streets/alleys accessible for a few minutes walking distance from the houses. This principle enabled the mosques to be one of the main public places of the neighbourhood insomuch as some of them blended into the public spaces of the adjacent streets and their courtyard acted as a Guzar (pass way) in connecting two lines of the streets/alleys. Tavassoli and Bonyadi (1992) conceive this as a prominent physical expression of the mosque as the house of God, the house which belongs to all people. For this, they present evidences from other Iranian cities, mainly concerning the *Jamia* mosques. In comparison, the city of *Sanandaj* contained this spatial principle in the mosques located in the neighbourhoods not just in the *Jamia* mosque so that one of the mosques called *Mezgat-e Dou-Derga* (the mosque of two gates) in the quarter of the ruling class. Considering this principle, the mosques' gates are often open to the people from early morning until late evening in *Sanandaj*. This was recognised by Ayazi (1992) as peculiar to the city of *Sannadaj* because in other Iranian cities the mosque is usually closed after each prayer time.[1]

Due to such frequency of meetings taking place in the mosques, continuous communication between the male populations could take place. Thus, the sense of community and social cohesion among the residents strengthened to such an extent that the people usually discussed their common problems concerning the neighbourhood at these daily gathering. The mosques and the marketplace were practically recognised as exclusive spaces which promoted "the quarter's social and spatial cohesion" (Madanipour, 2003: 212). Regarding that point, Rumi wrote that:

> the mosques were built to promote social integration and solidarity.
>
> That is the secret of why mosques were erected, so that the inhabitants of the parish might gather there and greater mercy and profit ensue. (Jalal al-Din Rumi, quoted in Madanipour, 2003: 210)

Such sense of community under the umbrella of common religious beliefs together with the socio-cultural circumstances of the Kurdish context promoted the social relationships among the residents. In this respect, Haqshenas (1999), who recounted his life history in poetry and was one of the interviewees, wrote that, there was a great trust among the neighbours to such an extent that the gates of their houses were often open for their easy socialisation. This means a great sense of collaboration and togetherness which also responded to the lack of adequate

[1] The Shiite people usually gather at the mosque three times a day not five times because they often pray early morning (*Fajr*), *Zuhr* (noon) and *Maghrib* (sunset). Two other times (*Asr*/afternoon and *Isha*/late evening) usually are combined with *Zuhr* and *Maghrib*.

laws and regulations for the security of its inhabitants (Sultanzadeh, 1986). In another view, it can be claimed that the compact structure of the neighbourhood's built form with narrow and curved lines (circulation system) and with minimum open public spaces, in a sense, are the result of such cultural values of togetherness. This is what recognised by Rainer (1977: 206) as the "human scale" of built form which presents an element of "livableness and neighbourhood". That is in contrast to the view of Southall (1998: 229) who wrote that, "irregular shaped quarters and their densely packed form took their basic solidarity from pervasive violence and insecurity which threatened everyone and strengthened the ties between them". Note that, the evidence presented in the previous chapter showed that the city comprised neighbourhoods with differing layouts, from the complicated form in the neighbourhood of bazaar (Aqa-zaman) to the rectangular form in the *Char-baq*. The morphological features of security defined in the neighbourhood of bazaar were mainly due to the social background of its inhabitants which was made up of the city's religious minorities. Equally, the War between the neighbourhoods mentioned earlier was not based on a religious theme like that of other Iranian cities. Here, the conflict was usually stimulated by political disputes that happened between the *Ardalan and Wakil*'s families settled in different neighbourhoods.

The convergence of these elements (mosque, *Hammam* and the houses of *Ulama*) along the principal road, created a particular spiritual dimension along the course of the *Bazaarche*. In addition to the occasions of daily prayers in unifying the people towards the central mosque by the first call of the muezzin, the spiritual dimension also had a strong impact on the consolidation of the people to respect each other on the occasions of birth, marriage, death and religious and national festivals (Ayazi, 1992). The importance of the mosque's function as a main stimulus to the people's gathering and their togetherness as a unified community can be seen in the following words from *Sheikh Ebrahim.*[2]

> During each daily prayer, the mosque of *Tuba-khanoum* (the central mosque of the neighbourhood located in the linear *Bazaarche*) was

---

[2] Interviewed (17.06.2004).

> crowded by the young and adults of the neighbourhood. On such occasions, the unity of the people was to such an extent that if somebody was absent among us we usually asked somebody else to call on him in his house and check the reason for non-attendance as we were used to seeing and to being aware of each other's circumstances in the daily prayers. Apart from this, we usually discussed the affairs of the neighbourhood. For example, if somebody had a problem and his close neighbour was aware of it, the neighbour brought up the case and we tried to solve the problem by recommendation of the Imam or other elders of the neighbourhood. Our regular discussions were mainly about the poor people and how to help them by gathering donations. More importantly, if somebody was holding a ceremony for his daughter or son (the case of marriage), his close neighbours usually consulted the issue with the Imam and then he announced it to the people for any help and collaboration.

Although the above descriptions predominantly dealt with the potentials of the linear *Bazaarche* along the principal road, the *Wakil's Bazaarche* was also contained the religious element of the mosque and the *Hammam*. But the point of difference is in relation to the spiritual atmosphere as this *Wakil's Bazaarche* does not show any evidence of such dimension due to the lack of *Ulama*'s houses in its environs (Fig. 5.1). The public function of those spatial elements mentioned above alongside the retail shops facilitated a base for the urban public elements of Chaikhana/teahouse,[3] as a key secular element in Kurdish social life, which the following section aims to explain.

Secular Elements

In the second context, it is possible to consider informal gathering within the houses and Chaikhana (teahouse) and their active roles in the social life of the neighbourhood, especially among adults and young people. More importantly, as Levkowitz (2018: 26) says, the Chaikhana serves as "the bedrock of public interaction in Kurdish society", particularly as the men's social realm. Before moving on to review of each element in

[3] It was usually called Qahwa-khanh (coffee house) in Sanandaj, but as it mainly serves tea instead of coffee I have put the traditional name of chaykhana for this entry.

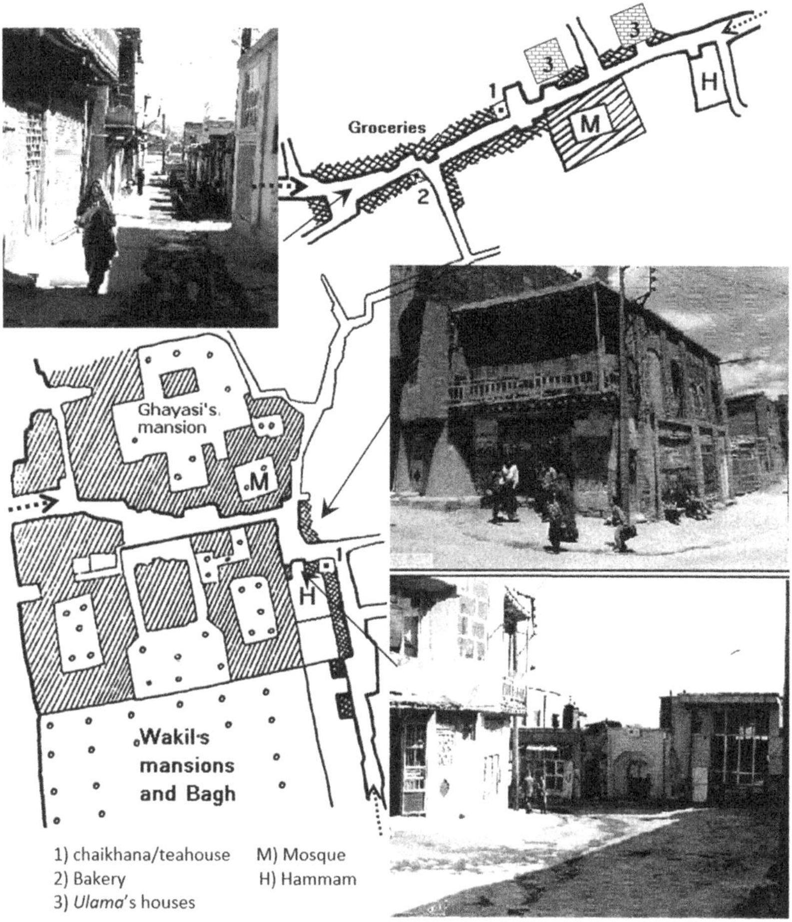

Fig. 5.1 The *Qatar-chyan linear Bazaarche* and its relation with the secondary centre, *Wakil's Bazaarche*. These two centres are the focal points of activities in the neighbourhood where anyone can go there freely (*Source* Images and map by author August 2003–2004)

this context, it is necessary to examine the notion of social background which contextualised these elements in Kurdish culture.

As discussed in the previous chapter, the Kurdish context showed evidence of some distinctive characteristics with respect to the sense of neighbourhood and social cohesion among its inhabitants. Although such a unified community structure was recognised by Muslim and non-Muslims scholars as flowing from the lessons of Islam extracted from two main sources, the *Quran* and *Sunnah*, in the case of the Kurdish context, apart from these two great sources of the knowledge, the unity must also be conceived from their socio-political and environmental circumstances formed along the history in mountainous area of Zagros mountains (Gunter, 2018). In explaining these circumstances, it is necessary to look at history, especially when the Median Empire, as the first unified Iranian government, was recognised as the main descendent of the Kurdish people (Gunter, 2018; Izady, 1992). It was founded in 728 BC and lasted until 550 BC. The importance of this empire is such that the Kurds usually link their background to it, and base their calendar and main national ceremony on the date of 612 BC (Kirmanj, 2014). This marks the date that their national existence was first recognised.

The important point of history is in the excluding of the Kurdish people from the government after the Median Empire and, consequently, pushing them to take refuge in the mountain areas of *Zagros*. This situation, with the harsh conditions of the mountains and creation of other governments in surrounding lands without a key role for Kurds, forced Kurdish society to become more isolated thus strengthening their internal solidarity and acting in a unified way in order to ease the circumstances created. This means that a strong sense of community was locally evoked by the determinants of nature and political power leading to a particular conceptual framework of socio-spatial structure which materialised the space in accordance with the cultural requirements of the inhabitants.

Considering the above review and the geographical setting of the Kurdish settlements in the rough mountainous region of Zagros, one can realise the pattern of social life in the Kurdish context. This geopolitical setting necessitated a kind of dense pattern of settlements, the piling of buildings upon each other, in order to optimise the restricted space. This pattern structured the place in a stepped configuration so that "one never knows whether one is standing on a floor or on a roof, since the terrace or open space which contains the floor of one house forms the roof of the one below", the roofs that are indeed the only open spaces of the settlement (Hansen, 1960: 30) (Fig. 5.2). In such a spatial structure,

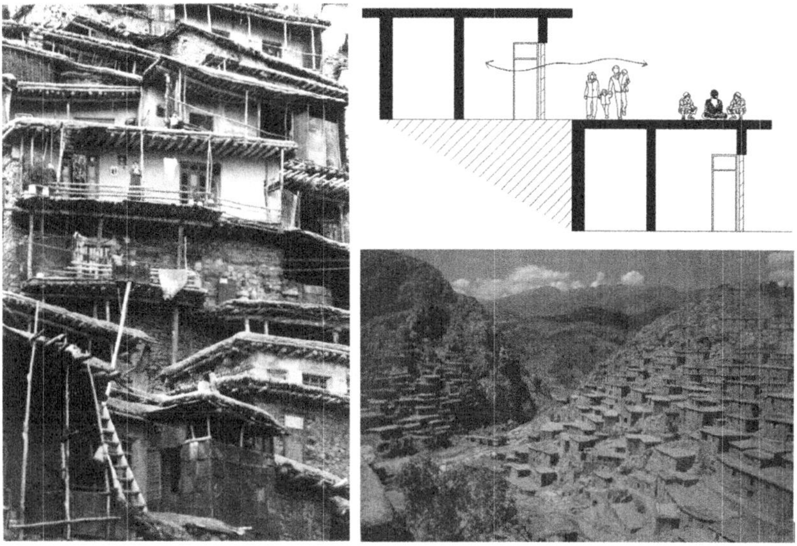

Fig. 5.2 The concept of neighbourhood in Kurdish society. Piling houses on each other in two settlements: first the village of *Tangesar* to the south west of *Sanandaj*, second *Palangan* one of the mound cities (*Sources* Above sketch by author, August 2005, left and right photos from collection of Mr. Gazarani)

social life is usually organised in close relation to the houses due to the form of the landscape. This means that the house was integrated into the domain of public life by inviting the neighbours to take leisure time in the open space in front of the house, the terrace platform which gives the inhabitants the ability to view the distant scenery.[4] In the words of Barth (1953: 105), who refers to this type of informal gathering as "Rooftop society", this gathering constantly formed a foundation for socialisation among the men of lower class status because of the friendly discussion among them, a form of interaction which is not suited to the status of nobles.[5]

4 This concept was transferred/redefined into the structure of houses in the city in the form of a veranda in front of the main part of the house flanked by two rooms overlooking the courtyard in front and the distant landscape. This defines the common plan of the houses called triplex-parts in the city.

5 It was called friendly discussion because as Barth (1953: 105) clarifies, "there is a tendency for the individuals of the highest prestige to stay away from such groups, since the joking

While the open space is a part of the house, it is also the only way for people to move around and at the same time, it is the rooftop of the house below. Although the house behind the open space has a boundary to define the territory of the family, in such intermingling functions it is also open towards the space before it the space, which is to some extent defined as a semi-extended part of the house. Due to such spatial structure, the openings of the houses facing the terraced platform were not fixed above eye level because the passer-by is not a stranger, his identity can be characterised as a neighbour with intimate character. In other words, it means that such houses are more intimate with the space before them so that they were not completely cut off from outdoor space and remain in communication. This "imparts a sense of fluidity between the inner and outer order" spaces (Baskaya & Symes, 1994: 149).

Accordingly, the coexistence described above alongside a long period of historical challenges with surrounding lands has helped mould the Kurdish culture and created strong social cohesion among its people in the isolated settlements within the *Zagros Mountains* so that they acted as an extended family to each other. More importantly, it maintained a pattern of visiting among the neighbours such that the house became an arena for some functions of the public life, especially a gathering place for neighbours to spend their leisure times during the evening. For example, Claudius James Rich (quoted in Barth, 1953: 103), who visited the city of *Sulaimani* in 1820, "was surprised at the regularities of the patterns of visiting, at the large groups that would congregate in the house of some or in the open, and spend their time smoking and drinking tea, while talking way into the small hours". As he continues, this regular gathering in connection with the feast occupies a very prominent place in the communal life of the people. It strengthened their solidarity and unity, as well as being a symbol of such unity. In addition, it also strengthened the status of hospitality among Kurdish people, leading to the institution of the guesthouse (*Mewan-khana* or *Dewa-khan*) (Leach, 1940) as the key element of the Kurdish social life. This institution played an

---

and horse-play may be rather personal and quite free, and the rules of proper behaviour easily forgotten".

important role in the nurture of Kurdish culture and its transmission between generations (Hassanpour, 1996; Kreyenbroek, 1996).

> Every traveller passing through a village can make a claim to the proverbial Kurdish hospitality. (Bruinessen, 1978: 82)

Mewan-khana (Guest House) or Dewan-khan (Court)

> In some guest-houses the chief and village elders would gather round the fire within a few feet of us and discuss us. This is most tantalising, as one can never quite hear what they are saying. (Hay, 1921: 55)

Based on the words of Bruinessen (1978: 83), the *Mewan-khana* was a place of rest for travellers within which they are entertained, given tea and a good meal, and a bed for the night. Apart from this, it was also the main place of male socialisation where all male villagers came and sat in the evening, and discussed daily matters before the head of village, "the actions of the various members of the group were coordinated and opinions and plans crystallised", decisions were taken, the young were taught traditions and etiquette and entertainment was centralised there (Barth, 1953: 104). For these reasons, sometimes it was called the *Dewan-khan* (court) or smoke-filled room. Apart from these, while the people were entertained with tea and cigarettes, singing and storytelling were performed to recall past events, recent events recounted by the guest were discussed, wandering preachers were listened to, and generally an important part of local entertainment took place (Allison, 1996). The later activities usually happened in the absence of the village's head man because the atmosphere of the *Mewan-khana* became a friendly club. Although entertaining guests and travellers costed people, they were the carriers of news which benefited the people by exchanging information and gossips. *Mewan-khana* was usually managed by the head of settlements who acted as the chairman of those formal gatherings. But generally, considering the social activities attributed to the *Mewan-khana,* Hay (1921: 49) perceived it as "more of a village club than the private property of the headman".

The mountainous feature of Kurdistan region, which formed an isolated pattern of settlements with few external connections, alongside the harsh weather of winter contextualised such spatial element to ease the conditions for travellers crossing the region and to facilitate communications between those settlements. The importance of *Dewa-khan* in the social life of Kurdish people, especially as a symbol of hospitality, was such that rural settlements maintained this element in the spatial structure of the houses, where one of the rooms, usually on the second floor and the best part of the house, was called the *Dewa-khan* in respect of the guest and used mostly for the entertainment of the guests.

> Hospitality is one of the finest features in the Kurds character. It is, it is true, enjoined by their religion, and the same custom prevails amongst other *Muhammadan* areas. But the Kurd has carried it to a fine art. (Hay, 1921: 49)

Chaikhana (Teahouse)

From the aforementioned points concerning the concept of *Dewa-khan*, it can be claimed that the function of Chaikhana mainly rooted from the *Dewa-khan's* function. The convincing evidence of this became clear when interviewees were asked about the leisure times in the past life of the city, most of them directly referred to the function of Chaikhana and its role in the social life of the people. This claim does not mean that the concept of the Chaikhana in other Iranian cities had no impact on this urban element in *Sanandaj*. My argument is that the main concept has a Kurdish background which borrowed some features from other Iranian cities.[6] For example, one can draw attention to the disposition of *Pahlavani* (championship) and *Lutigari* (generosity) as Iranian features which were mostly performed in Tehran's *Chaikhanaha* at the end of nineteenth century (Alemi, 1991), and became popular in *Sanandaj* as well.

In the *Qatar-chyan*, as Ayazi (1992) and Haqshenas (1999) state, it contained two active places of Chaikhana within which adults and

[6] In the Iranian context, the function of Chaikhana was evidenced since the period of *Safavids* times when it reached the height of their popularity in the *Qajar* times (Alemi, 1991).

young people, even from other parts of the city, passed their leisure time drinking tea, and talking about the events of the day or listening to a recitation of the poems from the *Shah-Name*, especially the drama of *Rustam* and *Suhrab*, by a story teller called *Naqqal* or *Morshed*. Of these *Chaikhanaha*, the one in the *Wakil's Bazaarche* was more popular for young people and *Pahlavanan* (champions) to gather there (Ayazi, 1992). As explained earlier, this character distinguishes it from the linear *Bazaarche* because the lack of *Ulama*'s houses in the environs of *Wakil's Bazaarche* made it more appropriate for secular than spiritual activities.

Similar to the function of *Dewa-khan*, long Autumn–winter evenings were the most popular time for the Chaikhana's function to gather people inside to spend their leisure times in a friendly atmosphere. Due to the presence of children and their desires for the continuation of the story, as Ayazi (1992) explained, the *Naqqal* had to continue his story till midnight. Apart from the performance of the *Naqqalli*, which contained important lessons for the children, such gatherings also strengthened the sense of unity among the neighbours by directing some of the debate to the problems of the residents and ways of dealing with them. More importantly, the *Chaikhanaha* were more active during the holy month of Ramadan. As Ayazi (1992) indicates, during this month, *Chaikhanaha* became the arena of playing *Turna*,[7] especially after the *Isha* (evening) and *Tarawih* prayers.[8]

Generally speaking, the social life of the neighbourhood, can be imagined from the following words of Mojtahedi[9] (one of the leading *Ulama*):

> After the *Fajr* prayers, we usually witnessed the arrival of some small caravans into the *Bazaarche* from the rural areas carrying the daily and weekly needs of the people. On this occasion, the *Kasaba* (traders) usually rushed to their shops to unload the mules they had ordered. From this

[7] *Turna* is a kind of game mostly between youths in which they play with matchboxes to represent the king and vizier (minister). The vizier usually is in charge of the *Turrna* (a piece of cloth made to be like a lash) to carry out the sentences of the king.

[8] This prayer is specific to the month of Ramadan which comprises 20 units of praying and usually lasted until shortly before the Fajr (daybreak).

[9] Interviewed 23.08.2004.

time onwards, the unloaded goods were arranged in front of the shops for customers. The *Bazaarche* became crowded with the coming and going of the people until the occasion of *Dzuhr* (noon) prayer. Although this praying created a pause in the tumult of the *Bazaarche*, it usually directed the stream of traders and other people towards the central mosque opposite to my house, where they all turned in one direction towards the *Qibla*. From afternoon onwards, the crowds in the *Bazaarche* were gradually diminished in favour of the *Chaikhanaha*'s function which often lasted until late night.

In summary, from the review above and based on the points conceptualised in the previous chapter, the principal road can be recognised as a key component in structuring the public places of the neighbourhood. Therefore, it is a unifying element as a place of movement and social intercourse not only in the level of neighbourhood but also across the whole city (Ayazi, 1992; Haqshenas, 1999).

Attitude to Visual Privacy

Following the notions concerning rooftop society and the Mewan-khana, which drew attention to the form of public life contextualisation in Kurdish society and mainly to the organisation of space, especially to the boundary between the two realms of public and private spaces, here it is important to examine the concept of privacy in Kurdish culture. This will help to explore the Kurdish culture attitudes to this concept in general and how it influenced the built form of the neighbourhood of *Qatar-chyan* in particular.

As seen in the literature, the main tension in that concept lay in the notion of sex segregation out of kin groups which was given careful attention in the organisation of space in Islamic cities. This was mainly highlighted by spatial elements in approaching the private domain from public thoroughfares and vice versa, the matter of overlooking and being overlooked, and the social norms concerning the presence of women in the public domain. As Al-Kodmany (2000) declares, such kind of strict boundary between the two realms has been put forward by the Islamic teaching as residential visual privacy is necessary for purity and goodness of life. Thus, it was recognised as a key factor in the organisation of space, and especially "the delineation between public and private spheres" across the residential quarters and even in the arrangement of space within the dwellings and its relationship with the space outside (Abu-Lughod, 1983; Madanipour, 1998; Memarian & Brown, 2003; Stewart, 2001: 177).

But in Kurdish culture, as explained, that tension was mitigated by the sense of neighbourhood and the effects of nature. Convincing evidence of this can be traced in the social behaviour and ceremonies of the Kurdish people. For example, the most popular Kurdish dance is called *Rash-balak*, which is usually performed in a circle with the young boys and girls taking each other's hands. In line with this notion, one can also look at the traditional dress of Kurdish women which reveals a contrast by using vivid colours, the colours that are in harmony with the natural landscape. Another important point is in relation to the status of women in Kurdish society which to some extent affected the mitigation of this tension.

Based on the accounts of orientalists' travellers, the status of Kurdish women in comparing with their neighbours was modest one as they enjoyed relative tolerance due to their socio-cultural background. As Galletti (2001: 209) indicates, "the western visitors have often described the strong character of the women whose role has always been relevant in Kurdish society as mother, partner, political chief and sometimes fighter and bandit". For example, in the words of Pietro della Valle (the well-known seventeenth-century Italian traveller quoted in Galletti, 2001: 210), women go about freely and unveiled and talk spontaneously with men both natives and foreigners, and entertain guests as head of the house. For Rich[10] (1836: 180–181, 85), "the Kurdish women are unveiled and freer than Arab and Turkish women". In the view of Nikitine (1956: 224), the Kurdish women usually socialise with men, to the extent that "they entertain the guests in absence of their husband and they are quite free and confident in their presence and do not cover their face as do the other Muslim women".

[10] He visited the region of Kurdistan in 1820. His image of Kurd and Kurdistan was recognised as the realistic and inquisitive approach to reality (because it was an internal image of the Kurdish society based on the lived experience (Mirawdeli, 2003).

In light of the above status, throughout Kurdish history, it is possible to find Kurdish women reaching high positions and becoming the political, in some cases even military, heads of tribes (Bruinessen, 2001; Nikitine, 1956) and famous poet and historiographer. In this regard, the most renowned are the three women mentioned by Sharaf-khan Bidlisi[11] in 1597, who came to power after the death of their husbands to pave the way for their sons to gain the power upon their adulthood. As Mojab declares, Bidlisi praises their ability to rule like the manner of males, and "calls one of them a lioness" (Mojab, 2003: 358). Another female ruler is the Khanzade Sultan, who ruled over two regions of Harrier and Soran to the east and north-east of Erbil (Iraqi Kurdistan) in the mid-seventeenth century in the time of Sultan Murad IV (1623–40). More importantly, she headed her army included 40,000–50,000 men so that several times raided into Iran (Grabolle-Çeliker, 2019). For more recent years, one can mention the name of Adela Khanum[12] of Halabja who gained the headship of the Jaf tribe in southern Kurdistan upon Usman Pasha's death in 1909 is the prominent figure in Kurdistan. She gained the title of Khan Bahadur (Prince of the heroes) from the British government due to her strong political character (Galletti, 2001). As Barth indicates, this was due to the prominent role she had as an administration assistance to Usman Pasha which enabled her to rule the region properly after the death of her husband till 1924 (Barth, 1953). In another case, based on the writings of Guest (Guest, 1993), Mayan Khatun (1874/1873–1957/1958) became a leader of Yezidis Emirate of Sheikhan (Iraqi Kurdistan—1913–1958/1957) when her older husband, Mir Ali Beg, died regardless of highly damaging rumors about her involvement in his death.[13] She gained such status in Yezidi society where "the clearly defined roles of the sexes" limit such range of women's activities (Allison, 2001).

---

[11] He wrote the first history of the Kurds, as a chronicle of Kurdish dynasties, mini-states, and principalities, in 1597.

[12] She was from the city of Sanandaj belonging to the aristocratic family of Wazieri, the main supporters of Ardalan emirate.

[13] Bruinessen (2001) titled such women as the Enterprising Lover due to the involvement of Mayan Khatun in love affair with a chieftain of the Muslim Doski tribe, Safr Agha.

From a religious perspective, it is possible to mention a spiritual leader of the Kurdish Jewry, Asenath/Osnat Barzani, who lived in the seventeenth century (1590–1670) in the Iraqi Kurdistan region (South Kurdistan) and became the head of a yeshiva after the death of her husband. Therefore, she was appointed as the head of the Yeshiva in Mosul, thus, became the first women leader of the entire community of the Jews of Kurdistan. She was considered the first and only female rabbi in the entirely of pre-modern Jewish history (Tallan & Taitz, 2009).

At the end of nineteenth century, it is possible to mention the name of a famous poet and historiographer known as Mastoureh Ardalan (1805–1848) who was born in Sanandaj and became the wife of Khasraw Khan Ardalan (1805–1834), the Wali of Ardalan Emirate, and migrated to Sulaimani after the death of her husband. She was recognised as the only-historiographer woman in the mid of nineteenth century in all the near and Middle East regions when many women were illiterate and were never permitted to go to maktabkhaneh (old schools of Iran) (Vasilyeva, 2000).

In line with the status of Kurdish women described above and the strong sense of neighbourhood among Kurdish people, we can claim that the spatial dimension of privacy in a Kurdish city is somewhat different from other Islamic cities. It means that the tension between the private and public spaces was not as strong as in the other Islamic cities to adopt more restriction on the presence of female in public spaces and even in defining more complicated ways of entering the houses, and ultimately to the restriction of the houses from being outward looking.

In *Qatar-chyan*, it is possible to observe the mitigation of visual privacy to such an extent that direct observation of the existing built form did not reveal extensive use of this concept in restricting the way of placing the entrance doors along the thoroughfares and windows over-looking each other. Considering these points, in many cases, the entrance doors were set just opposite each other along the narrow thoroughfares and even open to the view of passers-by from the thoroughfare (Fig. 5.7), and with windows overlooking each other and even overlooking the courtyard of other houses.

Following from the above concept, another point relates to the use of roof during the summer. As the extreme temperature could reach a maximum range of 30–40C in summer, the people usually used to spend the night on the house-top (Haqshenas, 1999). In this regard, although the dense pattern of houses aggregation and the topography of the city setting facilitated overlooking, there is no sign of any protecting wall on the roof-line of the houses. This point can be compared with other Iranian cities where the socio-cultural and environmental circumstances reinforced the feature of the built form requiring a wall around the roof-line to protect the realm of the private domain from any unexpected overlooking. The issue was discussed with the interviewees, they mostly emphasised on the good behaviour among the neighbours and the homogeneity of the people such that neighbours had priority over relatives. This point can be derived from the following words of Hairatsajadi[14]:

> Considering the sayings of Prophet Mohammad, the neighbours were *Mahram* (as close as relatives) to each other, as one of our sisters or brothers. Therefore, overlooking was no longer an important problem in our neighbourhood.

Or what, Sheikh Ebrahim[15] says:

> When someone was building or developing his house, the close neighbours would usually ask him to create an opening in the wall overlooking their house in order to give their family a sense of togetherness.

These circumstances, as Ayazi (1992) points out, had created a special potential in the social life of the neighbourhood to the extent that the roof was also another place of social intercourse between the neighbours on summer nights. While they socialised, they also shared their meals together.[16] Based on the design principles of Islamic cities pointed out by

---

[14] Interviewed 29.08.2004.

[15] Interviewed 17.09.2004.

[16] As Ayazi (1992: 336) explained, there was a custom among the people in the past life of the city which was named *Kasa-hawsa* (neighbourhood bowl). According to this custom, the people usually offered some parts of their meal to their neighbours. Continuation of this custom has left its mark on the interior features of the buildings. For him, three things were important when a house was going to be built: (1) creating a hatch or small window in the wall of the neighbour's house (2) the placing of a *Kursi* (a square table covered with quilts and

Hakim (1986), those notions above invade the visual privacy in Islamic law, whereas, they normally have to be prevented.

The claim above does not mean that there is no trace of visual privacy in the neighbourhood. The concept of privacy was applied rigorously in the houses of ruling class and rich people, and more so in the houses of those people that migrated or moved to Sanandaj from other Iranian cities because of their social background. Thus, they imitated the concept of *Maskan*[17] with the main focus of a courtyard structure and other related features.[18] From this perspective, there are different pattern houses across the neighbourhood from those with a rigid structure in respect of the privacy—with an introverted structure and a more formal presence in the world of public due to the socio-political status of its owner—to the more open pattern, with extroverted structure and more intimate with the world of *Kuche* (Alleys) by means of highlighting the feature of *Bar-haiwan* (open veranda)—the main device of house building in the region.

The above brief discussions of the concept of privacy allow us to move on to another relevant discussion concerning the women's presence in the public domain and its related spaces, where the level of social interaction can be explored.

blankets with a brazier under it to heat the legs and body during the winter) in rooms (3) and the placing of toilet at the corner of the *Housh* (courtyard). The first was in relation to the custom of meal sharing as it facilitated easy meal sharing among the close neighbours and their interactions.

[17] *Maskan* derives from the root *Sikun* which literally means Quiet (Warren & Fethi, 1982: 44).

[18] In line with this imitation, we can look at the entrance doors furnished with two splendid knockers: one a kind of mental ring for women, the other a large metal hummer for men. This was discussed with *Marufi* during our field visit, especially when we were observing the old city. He emphasised that, it is not from our culture, it is more an imitation from other Iranian cities because this high proportion of sex segregation was not of such importance for Kurdish people. The evidence can be traced through the settlements in rural areas of the region which lack these separated door knockers.

Women's Interactions in Public Domain

Islamic law requires that "sexual and emotional activities should be centred on the family nucleus to consolidate family life and reduce social stress", and avoids possible societal disorder (Edwards et al., 2006: 26; Memarian & Brown, 2003: 188). Public space is thus regarded as unsafe realm for Muslim women and better to be avoided. In these terms, the form of the city has been structured in such a way that it should minimise contact between men and women outside the kinship group through "tri-fold division of space" including private, semi-private/semi-public and public spaces (Abu-Lughod, 1983: 66). The logical consequence of such pattern originated from visual privacy is to create clear distinctions and hierarchical spaces to protect women from the eyes of strangers. For this reason, the private domain has acquired priority over public spaces as the home contains most aspects of everyday life in its courtyard structure, where women can enjoy the pleasures of life in the open air and in strict seclusion. Thus, preference is given to the inner rather than outer façades, with the outside façades normally being left blank with small openings above the eye level of the passer-by to prevent them from looking in. The only exception is the doorway which was usually decorated to indicate the habitation within.

The wealthy Moslem was able to ensure such segregation by spatial contextualisation of two social concepts of *Mahram* and *Na-mahram*[19] so that they regulated the thresholds of interaction between male and female inside the house. In such families, the female had no need to go out and communicate with the world of public spaces because everything was done for them by servants (Mazumdar & Mazumdar, 2001). On the other hand, the women of commoners had to go out for shopping and work in order to contribute to the income of their family. Hence, they had to interact with the public spaces of the city, especially the

---

[19] ***Mahram*:** People who are close, immediate family like brother, sister, father, mother, uncle, aunt, grandparents. *Mahram* people cannot marry each other (except husband and wife are married and Mahram). Thus, it is "the legal term denoting a relationship by blood, marriage or sexual union which makes marriage between persons so related forbidden" (Khatib-Chahidi, 1981: 114). ***Na-mahram*:** Male and female who are not close family are *Na-mahram* to each other.

centre of the neighbourhood. As Campanile (1818 quoted in Galletti, 2001: 210, 211) indicates, who lived in Kurdistan from 1802 to 1815, "the women's work was not limited to domestic chores". As he observed, "outside Mosul, the women were selling beautiful pearls".

There is a contradictory point concerning the unsafeness of public spaces for women and the notion of *Umma* (the community of believers and brotherhood) in Islamic discourse. Based on the Islamic teaching originated from Holy Qur'an, all men and women are brothers and sisters, especially when Islamic religion was recognised as a force of dynamic influence which provided the basis for social cohesion (Saleh, 2001).

> Believers are surely brothers. So, restore friendship among your brothers. Have fear of Allah so that you may attain mercy. (Qur'an, 49:10 quoted in Memarian and Brown [2003])

This is also in line with the organisation of neighbourhoods in Islamic cities as "geographical entities as well as homogeneous communities which were closely knit", where the notion of solidarity was strengthened by family, clientage, common village origin, common ethnic origin, common religious adherence and shared occupational ties (Madanipour, 1998: 237). This means a high degree of safety for female socialisation across the spatial structure of the neighbourhoods, especially its main centre of gravity. Another important notion relates to the necessity of the veil/*Hejab* for Muslim women by the verses of Holy *Quran*[20] which explains the possibility of women's presence and interactions in public spaces.

Women's Public Places in the Neighbourhood

Following the general points discussed earlier concerning the nature of Kurdish society regarding the status of women and their presence and

[20] "And tell the believing women to lower their gaze and be modest, and to display of their adornment only that which is apparent, and to draw their veils over their bosoms (*Sure*, ch.24, verses 30–31 quoted in Mazumdar and Mazumdar" [2001: 306]).

interaction in public spaces, here in this section the female's interaction in public spaces is discussed by referring to two neighbourhoods of *Jawr-awa* and *Qatar-chyan*. As aforementioned, the inhabitants of *Jawr-awa*, both male and female, were mostly poor and, thus, worked for the ruling class and wealthy people. It follows then the females had to go beyond the boundaries of their families, passing through different alleys/streets to reach there. On their way, they had to be in touch with people as well, particularly on their way to returning home they visited the neighbourhood's *Bazaarche* for daily shopping.

> I remember the period when we were living in our house in the *Jawr-awa*. As my father was one of the *Ulama* and teachers, we had rather a grand house, which apart from its main structure contained four areas at the front called *Zheer-mala* (a separate small house for a poor family) in the southeast corner of the courtyard, a stable and a charcoal store in the south, and *Sar-tanour* (the place of bakery) to the southwest of the house. For your information, those last three parts resemble an extension of the house towards the south. The structure of main part of the house comprised three parts with two corridors in between. Two or three times a week, two women usually came to our house from the *Jawr-awa* to bake bread in the bakery. As baking bread has a long process of preparation, especially kneading the dough, one of the women had to come in the early morning after *fajr* (daybreak).[21]

The above words of the interviewee, to some extent, clarify the circumstances of both the wealthy people and the commoners with respect to the presence of women in outdoor spaces. Most importantly, the working times of commoner women explain their presence early in the morning in the public spaces of the thoroughfares and even in the houses of other people for the sake of their families' income. The other point relates to the concept of *Zheer-mala* and its manifestation in the structure of the houses, especially the houses of wealthy people. It is somewhat peculiar to the city of *Sanandaj* and there is no similar concept in the Persian language. It is a real house alongside the main house but close to the

[21] Hairatsajadi (interviewed 29.08.2004) was one of the local researchers over 70 years old. His father was recognised as the founder of education system in *Sanandaj*. As one of the notables, his house was close to the quarter of ruling class.

entrance door. Generally speaking, it can be said that it is a concept lying between the two concepts of tenant and servant. This means that the status of *Zheer-mala* is not like a servant because it has a semi-independent structure so that he/she is not paid any money and they can have a job outside. Furthermore, this person is not a tenant because he/she has to do some unexpected business. Such persons might be invited to share in the meals of the host.

As a higher status family, I asked Mr. Hairatsajadi to make clear that how his mother usually socialised with the neighbours. He simply said that "my family was from notables and we usually dealt with those families whose status was the same as us. This does not mean that my mother had no relationships with the neighbours because they visited us for the sake of some business or to raise some questions concerning their own life or the neighbourhood". This defines a kind of formal interaction with commoners.

In the *Qatar-chyan*, as the neighbourhood of muleteers, the circumstance is slightly different from the *Jawr-awa* because here the adult male had to leave his family for a long time to transport products from the region to other cities. Hence, the adult female was responsible for many socio-economic affairs of the family including attending public spaces of the neighbourhood for shopping.

> On the occasion of journey to other cities (mostly *Hamadan*, *Zanjan*, *Rasht* ...), those people in charge of transportation affairs usually put their family under the care of one of the neighbours. The neighbour looked after them as his relatives. He supplied all their shopping apart from the unexpected items which were bought by the adult female of the house from the closest *Bazaarche*. (Sheikh Ebrahim)[22]
>
> In the past, the social relationship of the people was not bound up to the economic issues as today. The neighbours had a good relationship with each other and looking after a neighbour was a rather common practice among them. In line with this, the entrance doors were usually open for easy movement between them and each house had an opening to the adjacent one to be used for meal exchange. In these circumstances, those families whose men were away to other cities for the livelihood

[22] Interviewed 17.09.2004.

> of their families, their housework was usually managed by their closest neighbours or adult females. (Mujtahedi)[23]

From the points mentioned, the notion of unsafeness of public spaces for women's presence in Islamic cities raised by orientalists and some Islamic scholars can be recognised as a narrow view which was only based on the life of wealthy people and ignored the challenges of life in other layers of the community.

Apart from thoroughfares and Bazarcheha, where both women and men can attend in the level of the neighbourhood, there are other places which were only used by women within the neighbourhood and even across the city. They are *Hammam*, cemeteries, shrines and tombs of saints, *Kanni* (springs) and *Bar-mal* (doorstep). These places will be examined in the following sections.

Hammam (Public Bathhouse)

The *Hammaam* is another common feature of Islamic cities within which "health and social institutions" were located (Kheirabadi, 1991) in both city and neighbourhood levels. As Ayazi (1992) explains, the *Hammams* were used according to a timetable suited to the men and women, so that they were usually assigned for women's use twice a week, Saturday and Tuesday.[24] As he continues, apart from its function for washing and personal cleansing, the *Hammam* was an important meeting place where they usually stayed for a long time in a large heated space together with an abundant supply of hot water, often from 9 o'clock to sunset. Apart from washing, they also eat and chat together in getting and spreading the news and gossips about each other. From this point, it was recognised as a main place of information exchange, bonding friendships and mate selection. There is a proverb which says: "Use the bath water to make

[23] Interviewed 23.08.2004.

[24] As he cited from his forebears, there was a different way of managing the timetable of *Hammam*. It was usually used by men from midnight until 9 o'clock and from that time onwards it was the turn of women. At 9 o'clock, the custom was that a person blew a horn on the roof of the *Hammam* to inform both men and women that the time of *Hammâm* should be changed (Ayazi, 1992: 410).

friends" which means that only by pouring pitchers of water over each others' shoulders could acquaintance and friendship be gained (Bamdad quoted in Mazumdar and Mazumdar [2001: 308]).

The neighbourhood contained two places of *Hammams*; the *Tuba-khanoum*'s *Hammmm* along the principal road in the linear *Bazaarch* and the *Wakil's Hammm* in the *Wakil's Bazaarche* adjacent to the spatial element of *Qahve-khane* (Fig. 5.1). The spatial structure of the *Hammams* comprises washing and dressing rooms as well as other services.

Cemeteries

Another place for women to gather is in the cemeteries. Based on religious beliefs, people usually gather in the cemeteries in the late afternoon on each Monday and Thursday to read prayers to their dead, and to re-establish their feelings for them. As it is usually frequented by women, it can be recognised as a place where they meet each other and they can talk about the events and issues of their life. As the city was formed based on the identifiable urban neighbourhoods and each neighbourhood had its own cemetery close by, this facilitated the easy movement of women to visit the cemeteries.

Shrines/Tombs of Saints and *Takya/Khanegha*

The shrines and tombs of Muslim saints were given an important position in the Iranian context due to the cultural beliefs echoed by Shiite ideology. From the viewpoint of psychology, women participate more actively than men in *Ziyarat*/visits to these places (Fruzzetti 1980 cited in Mazumdar and Mazumdar [2001]). Although the people of *Sanandaj* mostly adhere to Sunni beliefs, the *Ziyrat* of saints was to some extent the custom due to their long shared cultural background with other Iranian cities. There are two main shrines in the city; one in the *Meyan-qala* called the place of *Peer-Omar*, and the other in the *Jawr-awa* named the shrine of *Hajara-khatoon*. Apart from these, there are other small shrines within each neighbourhood which were usually used by women to express their inadequacies, failures, problems, or request favours and pay/perform their *Nazar* (vows) (Fig. 5.3). Another place of women's gathering is the houses of *Masheikh* or *Sddat* (descendants of Imams). In these places, they usually ask the saint to help them to resolve their family problems. Such houses usually consist of a waiting room where the women can chat with each other before and after meeting with the *Sheikh* or *Saddat*.

Following this category, women also attend some religious ceremonies as the members of a sect of Dervishes. Here, the women and the men participate in praise of God, but in separate rooms. Such religious ceremonies usually happen in places called *Takya* or *Khanegha* centres of spiritual training. Within *Qatar-chyan*, there is no particular building called *Takya* or *Khanegha*. For this reason, sometimes the mosques and houses instead were used for the relevant ceremonies. Or they had to travel further away to other neighbourhoods to participate in the ceremony of interested order, *Qadiriya* and *Naqshbandiya*. The lack of old buildings of *Takya* or *Khanegha* in the neighbourhood and even in the city can be related to the far more recent existence of this phenomenon

Fig. 5.3 Left, the cemetery of Sheikh Mohammad Baqer in the west part of the neighbourhood, which contained the shrine of this sheikh on the top of the hill (Bahrami, 2001). Right, the women's gathering place beside one of the shrines in the cemetery of Sheikhan. The shrine is located in the middle part of the street due to the physical development of the city. While the cars are driving by on both sides it still functions as a place for women's social interaction (author August 2003)

in Kurdistan, which dates back to the first half of nineteenth century, and the fact that they were spread more among the peasants in the countryside than the townsmen (Bruinessen, 1978).

Kaani (Spring)

In addition to the above public places which are also common in other Islamic cities, the site of *Kaani*/spring is to some extent peculiar to the Kurdish society. The importance of the *Kaani* in women's social life can be compared with the function of *Dewa-khan* and Chaikhana/*Qahve-khana* in male realm.

The *Kaani* is where women usually filled pitchers and carried them on their shoulders to supply drinking water to their families or for others as a way of earning money. Due to the daily need for water drinking, it can be recognised as a frequent gathering place for women and the one that had a major role in their socialisation. Its importance in the life of Kurdish women was such that Marufi[25] refers to it as *Kabul-akhbar*, the news centre or the place of knowledge exchange. He goes on to say that fetching water was a pretext for them to visit the *kaani* frequently to meet others. Barth (1953: 106) found the water-hole, "where water for the house is fetched, and clothes washed and beaten", as the main place of women aggregations where gossips are exchanged.

> Apart from collecting water for our family, it was really fun to go there and meet other women from different parts of the neighbourhood and, more importantly, have access to information which the men did not havc. (Droudgar)[26]

Based on the words of Ayazi (1992: 420–423), there were three sources of water supply in the city: *Qanat* (subterranean canals), which supplied the needs of public places like mosques, *Takya*, *Hammam,* the houses of nobles and some of the wealthy people, and the well and *Kaani*/springs which used by other people, mostly commoners. The well

[25] Interviewed 22.10.2004.

[26] Interviewed 11.08.2004.

water was usually used for washing dishes and clothes, irrigation and similar purposes. Because of health risks in using well water for drinking, and due to the location of the city in the intermountain basin, springs were mostly used to supply the daily drinking needs of the people. Due to its location in intermountain basin, the city contained more than 25 springs. Of this number, the *Qatar-chyan* had four springs mostly in the valley close to the place of the *Qula mosque*, in the first phase of neighbourhood formation.

As most of the city springs were destroyed by the physical development of the city, their function as a gathering place for women must be traced from evidences available in the rural areas, where drinking water for people was usually supplied by one or two springs in the village. In this regard, my own experiences, as I lived in one of the villages in Kurdistan called *Qala-kouna* during my primary school years, are used. The experiences mentioned were in a small village to the far north of *Sanandaj*, which contained approximately 500 people. The village's drinking water was supplied by a single spring in the south part of the village, where it was connected to a subterranean canal at the end of the settlement. As household affairs were mostly the responsibility of the women, the drinking water was supplied by them from that spring. For this reason, they frequently had to visit the spring during the day. The peak time for these visits was evening when the men were coming back from outdoor work on farms. Generally speaking, the importance of this custom was such that there are many folk songs and music concerning the course of a young woman's journey from her house to the spring carrying a pitcher of water and a young man, who desires a sip of water from the pitcher.

> It is evening time again, so pick up the jug madam, while the spring's walk way is waiting for your appearance. (poet: Ahmad Hardi)[27]

Similar notions can be found in other Islamic countries, especially in the Middle East region. But the point of difference is in the way women are present in the public spaces, on the way to and from the spring and

[27] He was born in 1922 in Sulaimani, Southern Kurdistan, his first and only book of poems, the Secret of Solitude, was published in 1957 (Hardi, 2004).

even their gathering close to the mouse of spring. For example, Hansen (1960: 85) describes the following practices around washing places in a Shiite village in Bahrain:

> The washing places were sufficiently protected by means of a low semi-circular wall, or by the natural rise of the ground, for the women to work by the water without being obliged to veil themselves. Furthermore, neither men nor youths appeared in the vicinity of the washing place. On the other hand, the women were very careful to be veiled on their way to and from the washing place.

In contrast, in the Kurdish villages of Kurdistan and even in the past life of the city—especially in the *Qatar-chyan* neighborhood, the signs of protection were not in evidence for women's presence at the place *Kaani*. Furthermore, on their way to and from the place spring, they were not obliged to veil (*Chador*) themselves.

Although the above descriptions may not completely match the function of *Kaani* in the neighbourhood, it can be recognised as a partial knowledge concerning the function of this spatial element in the social life of Kurdish women due to the deeply rooted links between the villages and the city (Vasilyeva, 2000). In line with this point, Ayazi (1992: 423) mentions the water carriers/women who were professional in this work and paid monthly. Moreover, Rangavary[28] (1999) explains that the area around springs usually was crowded with women who were filling their containers and chatting with each other as well.

Bar-Mal (Doorstep)

The last place of women's social interaction is the place of *Bar-mal* (doorstep). This is in line with the second context of social life in the Kurdish society. It was a place usually frequented by women each morning after finishing their duties at home (Marufi[29]), particularly when the man left the house and the arena of the public domain was

[28] Tape recorded interview in 1999.

[29] Interview 20.10.2004.

appropriated for women to socialise with their close neighbours. While they are chatting with each other, their children play freely around (Fig. 5.4).

Apart from spending leisure time with neighbours, such place of social gathering was a good opportunity for commoners' women to engage with the manual wool-spinning as it was also a common activity among the Kurdish women. This activity was usually performed outside in the open spaces in front of the entrance door because some parts of it required other women's help. More importantly, being outside helped them to minimise any health problems related to their lungs and to keep the house clean. Thus, by doing this, on the one hand, they increased the income of their families either by selling wool directly or indirectly by using it to weave rugs for sale at the local market and, on the other hand, they gave themselves an opportunity to socialise with other women.

Based on the local opinions, there are two different views of the women gathering in front of the house. In the view of those that relate themselves to the wealthy and *Ulama* families, this form of

Fig. 5.4 The Kaani and Bar-mal as important places of women's gathering and their socialisation (*Source* Author August, 2003–2004)

gathering took place mostly in the house or in the corridor area just behind the entrance door which contained the place of *Sar-tanour*, the bakery (Hairatsajadi[30]; Mujtahedi[31]; Sadeq-wazeri).[32] The *Sar-tanour*, according to their views, was an important place for women to gather. While some were baking bread, others gathered around and talked with each other. For these higher status families, due to both the amount of space in the house and the values and lifestyle that they regarded for themselves, women sitting in *Bar-mal* were not seen as good manners and it was regarded as a custom more suited to the realm of commoners' life. Nowadays, this custom is also practiced by them in many parts of the city, particularly in informal settlements.

> I have never sat in the bar-mal because it was not considered as good behaviour in my family. I usually socialise with other women in the house. (Sadeq-wazeri)[33]

For other interviewees (Marufi,[34]Sheikh Abrahim[35]), sitting in the *Bar-mal*'s was a custom among the women of the neighbourhood where they would spend their leisure times and even do some household work, mostly wool-spinning.

> I remember that my mother usually spent her leisure time talking with the neighbours in front of the house, especially when my father left the house for his daily business. Most importantly, when he came back, they ended the meeting in respect of my father. This respect was not just for my father, it was mostly for elderly people due to the modest behaviour of women in the neighbourhood. (Sheikh Ebrahim)

From the above section, it can be realised that the life of women in Sanandaj is not confined solely to the private domain. But their religion,

---

30 Interviewed 28.05.2004.

31 Interviewed 23.08.2004.

32 Interviewed 21.08.2004.

33 Interviewed 19.09.2004.

34 interviewed 20.10.2004.

35 interviewed 17.09.2004.

as a way of life, and their cultural background rooted in circumstances peculiar to the place and its history provided them with important social roles and enabled them to go out of the house and be in touch with different arena of public life.

Street Pattern

After analysing the social factors underlying the concept of public domain in the two realms of men and women, it is time to move on to the spatial features of the neighbourhood that result from those social factors and the features of the city's setting. This section examines the street pattern of the *Qatar-chyan* and related features that are defined as experiential qualities of space.

As explained, the principal road functioned as the main artery and collector in merging the traffic lines of the neighbourhood towards the city centre and also from the other neighbourhoods towards the recreational spaces on the slopes of the *Awear Mountains*. Along this road, the neighbourhood's main *Bazaarche* formed in which different social activities evolved. Due to such prominent role, it also influenced the overall structure of the street pattern of the neighbourhood so that the primary thoroughfares mainly branched out from it towards the two parts of the quarter. These thoroughfares as arms of the principal route have fed the secondary streets (*Kucheha*), which circled the boundary of clustered houses, where confined areas of courtyard houses are located. From this point, the spatial structure of the quarter can be recognised as a unified graph so that the other streets were arranged around the principal route in a relatively hierarchical system of distributor, feeder and collector routes (Fig. 5.5).

The last part of the above pattern, as it does not have an outlet and continuation, is recognised as the *Bunbast/cul-de-sac*, which defines the realm of semi-public/semi-private spaces among close neighbours. This was recognised as a widespread form of street in the typical Islamic city, which characterises the sense of individuality and seclusion of private areas from thoroughfares among close neighbours (Monteguin, 1983). In some cases, it was owned in common by neighbours and also enclosed by

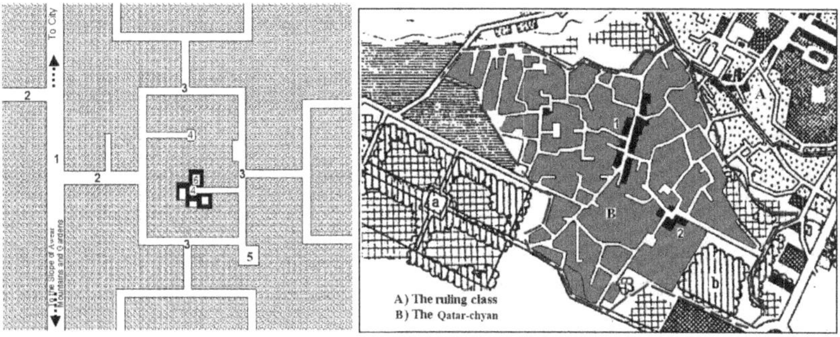

Fig. 5.5 The street pattern of the neighbourhood. (1) Principal road, (2) Primary *Kuche*, (3) Secondary *Kuche*, (4) *Bunbast*, (5) *Maidanche*, (6) Courtyard (*Sources* Left drawing by author, Autumn 2005 and right Map from Alten [1958])

a common gate, which was usually closed and barred at night. This character of the *Bunbast* was recognised as "a basic unit in the self-policing nature of the Islamic city" (Morris, 1994: 389). As it faced away from the realm of the public thoroughfare, this structure also created an intense silence and tranquillity for its inhabitants, especially for women's gatherings and their socialisation with neighbours. In some cases, it takes form of a *Maidanche* (a small square) set back from the thoroughfare (Fig. 5.5). This is a confined space surrounded by the nearby houses where women can sit in front of the entrance door and socialise with each other while their children are free.

In spite of the main principal road of and the street along the north-east part of garden and mansions of *Wakil*'s family, there are no feelings of a continual flux of traffic (a conduit of uninterrupted motion) as the sensation of flow is always broken by sharp angular bends and sudden shifts of directions in overall street circulation system of the neighbourhood.

To sum up, the street circulation system above symbolises and controls the transition from public through semi-public and semi-private areas to the private domain. It is a means of signalling "the changes of possession, of territory, of control and behaviour" (Lawson, 2001: 166).

Experiential Qualities of Space

In the process of managing the traditional neighbourhoods' affairs, building activities were usually determined based on a mutual respect between neighbours regarding access to breeze, sun and view[36,37] (Abu-Ghazzeh, 1999; Rainer, 1977). In such way of building, the circulation movement, between the houses, gained particular spatial diversity as it did not follow a preconceived plan due to the absence of any formal agents.

Due to the social structure of the neighbourhood, the coexistence of rich and poor, the passer-by can experience diversity of houses' appearances and different attitudes to manifest identity, so that his eyes never become tired because in each sequence he can view a new spatial quality (Rainer, 1977). Where the status of owner allowed, the house was distinguished by a long *Bar-haiwan*[38]—an open veranda facing the natural landscape or public domain—entrance porch, and ornamentation of the gateway (Fig. 5.6). Where high status is absent, the house was constructed with a simple appearance with a small *Bar-haiwan* but still in harmony with surrounding houses, which can be interpreted as modest houses "striving for dignity and respect" (Lombard, 2014: 16) (Fig. 5.6). In other words, this point expresses "solidarity with the neighbours", by using the same shape and structure but in its simplest way it offers, "a sort of tribute" to the community (Lawson, 2001: 36).

The above spatial qualities were enriched by maintaining a relationship with nature and distant scenery through the feature of a *Bar-haiwan*. This can be recognised as the quality of the terraced order in the mountain areas so that one can experience the distant and the close sceneries simultaneously. It indicates the presence of that place in this place. This

---

[36] According to Rainer (1977: 51), there was a tradition in Iranian town planning that, "you shall build your house such that you obtain sun and view, but you shall not take away sun and view from your neighbour".

[37] This was confirmed by the saying of the prophet: Do you know the rights of the neighbour... you must not build to exclude the breeze from him, unless you have his permission (Abu-Ghazzeh, 1994: 55).

[38] *Bar-haiwan* comprises two words: *Bar* means front and *Haiwan* is the right word for veranda—there is a similar word in Farsi—*Eyvan*. It means an open veranda in front of the main part of the house.

Fig. 5.6 Above: higher status houses with long *Bar-haiwan* and the entrance porch (*Sources* Author August 2004. Bottom: Modest houses with a small *Bar-haiwan* and simple decoration of the gateway; Author September 2018 and August 2003)

structure is not a prevailing feature in other Iranian cities located on the central plateau, where the features of the terrain did not support such a spatial pattern.

In the circulation system of the quarter, because of the social structure discussed earlier, the duality of place was maintained as an observer can experience both the blank houses with introverted structure shut off from the surrounding thoroughfares by high and solid walls, and outward-looking houses with the spatial feature of *Bar-haiwan* open to the public realm and surrounding sceneries (Fig. 5.6).

The visual quality of the surrounding scenery in some cases even influenced the houses of wealthy and notable's people whose houses were structured in line with privacy with central courtyard. In such houses, the builder maintained a long *Bar-haiwan* (in this case it is called *Khulam-gardesh*) facing the public thoroughfare on the second floor (Fig. 5.6). In addition, the walls of first floor facing public thoroughfares were not left blank as they contained rhythmic decoration mediating the form and quality features of the surrounding nature. However, this rhythmic decoration was not affordable by the commoners. From another point of view, such mingling of tight and open structures, simplicity and complexity can be defined as a "kind of hybrid form that make reference to different sources of ideas, levels of meaning" and, one may add, of status (Kellett, 2003: 96).

Another experiential quality of space along the circulation system of street is in relation to the houses' entry spaces. Entrances are recognised as one of the key features of built form because they are transition elements from one domain to another and communicate the role of occupants as well (Lawson, 2001; Moughtin, 2003). In the neighbourhood, the house's entrance doors show particular patterns in which the entrances were exposed to the view of passers-by in its circulation system (Fig. 5.7). In some cases, the width of the passageway narrowed due to the size of entrance door. This declares that the visual perspective was to some extent controlled according to the view of passers-by. In addition, due to

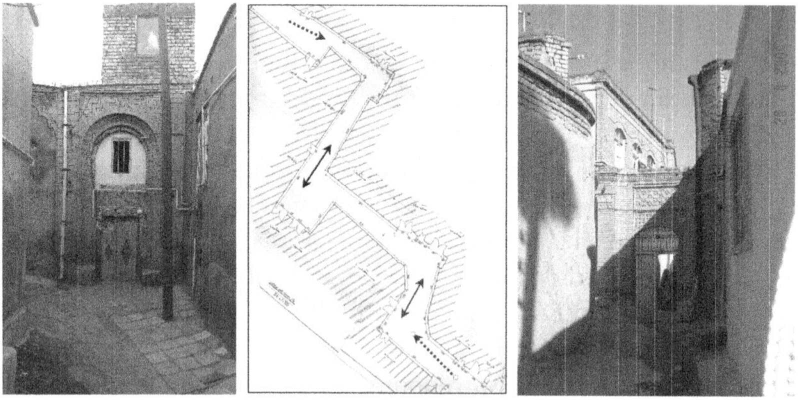

Fig. 5.7 Diversity of scale and portal decoration which created and enhanced visual qualities of the neighbourhood circulation network (*Sources* Author, right image, June 2019, drawing and left image August 2004)

the social structure of the neighbourhood where rich and poor are living side by side together, different images of entry spaces with a diversity of scale and portal decoration created and enhanced visual qualities of the neighbourhood circulation network (Fig. 5.7). This was recognised as a harmful view of a passer-by into the houses in the Islamic cities (Hakim, 1986).

Considering the structure and form of the entrance doors, two main periods are recognisable. The first period includes the time of *Qajar* dynasty from 1794 to 1873. Most of the registered houses by the Kurdistan Cultural Heritage Organization (KCHO) belong to this period. The second period comprises the end of *Qajar* dynasty in 1794 to the end of first Pahlavi dynasty in 1944. For the first period, the following features can be highlighted (Fig. 5.4, down photos on the middle and right positions):

- Elevated structure with a designed framework
- Two seating platforms (*Khaje-nesheen*) on both sides of the entrance door. Depending on the status of the owner, the size and number of platforms, niches and brick decorations might vary. For the visitor, especially in the houses of ruling class and notables, its construction is logical because it affords a place for him to rest while awaiting permission to enter the house or conduct other business (Fig. 6.3). In addition, depending on the location of the house along the thoroughfare, it was a popular place to spend leisure time where people could sit and talk to each other as well (Fig. 5.8) (Sultanzadeh, 1994).
- Semi-octagon space in front of the entrance door (Fig. 6.3 photos on the right position). This, with various sizes, functioned as a semi-public/semi-private and transitional space to the houses of wealthy and ruling classes. In some cases, it was covered with a semi-dome-shaped structure.

The second period was to some extent a result of the "first high-ranking contact with western culture [which] occurred when *Naser o-Din Shah* (1848–1896), the *Qajar* King, went to Paris and was enraptured by it" (Katouzian, 1996: 34). That was the beginning of modern thought in Iran which led to socio-economic changes and the introduction of

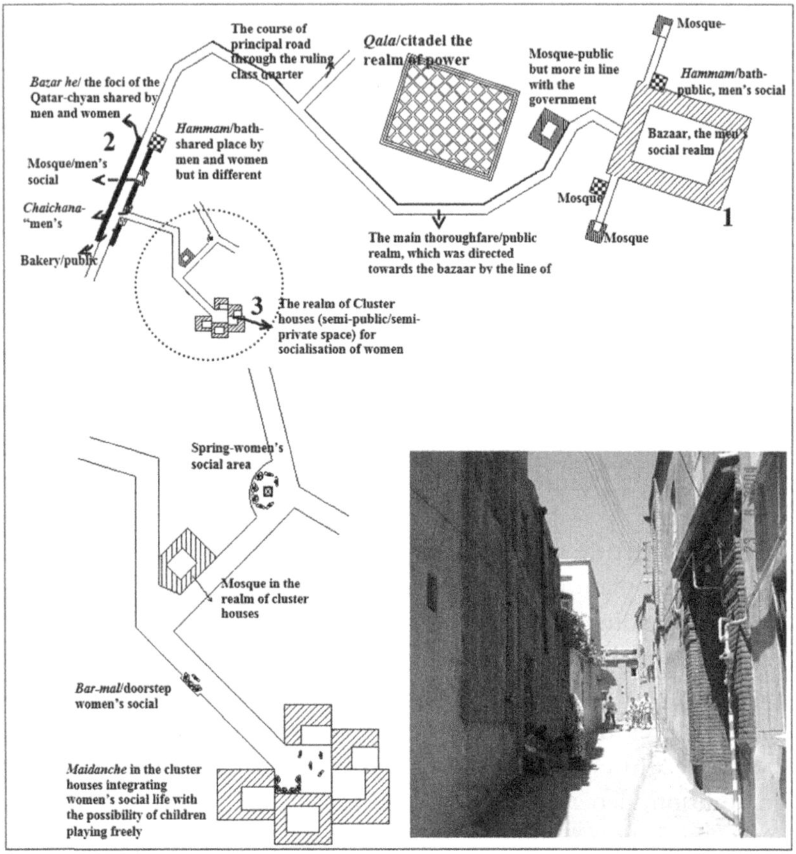

Fig. 5.8 Summary figure showing the socio-spatial dimension of the *Qatar-chyan* neighbourhood. The hierarchical movement from the bazaar to the cluster houses as the main domain of women's social life: (1) the bazaar complex as mainly in the domain of men's social life, (2) the *Bazaarche* as the shared domain, (3) cluster houses which contained the elements of *Maidanche*, spring and possibility of gathering by the doorstep (*Bar-mal*) and is mainly the domain of women's social interaction (*Source* Author August 2004)

new ideas to the structure of traditional society. It first happened in the capital city of Tehran and then echoed in other cities. These changes affected the overall structure of the entry spaces by introducing a new spatial element called *Khyaban* (Western Avenue). The presence of this element blurred the distinction between public, semi-public/semi-private and private realms because it neglected the tri-fold spaces in the circulation system of Iranian cities so that the buildings can stand immediately on the realm of the *Khyaban* without the transitional spaces of the traditional city. Due to these changes, the social and cultural functions of the entry spaces were neglected and, its size was reduced and other parts eliminated in favour of simple ornamentation of external parts of the buildings. More importantly, the spatial element of *Khaje-neshen* lost its function as a result of those changes in values and lifestyle. This trend also neglected the entry space as a distinct part of the house as it gave more importance to other parts to change the introverted structure of houses into extroverted ones to some extent similar to western styles. From these points, the features of this type are as follows (Fig. 5.7, right photo):

Small size of entry space in comparison with the elevation of the house
Lack of semi-octagon space with platforms and, thus, the presence of the doorway immediately beside the thoroughfare
Using different forms of doorway structures and distorting the original form of the semi-dome structure

Conclusion

This chapter examined the public domain in the level of neighbourhood considering further the matter of social life in Kurdish culture. The perceived socio-cultural background and the spatial features show that Kurdish culture is responsive to the concept of Genius Loci, and the dwellings in particular have a strong relationship to the landscape. The value of landscape somewhat influenced the structure of the dwellings mostly in indigenous architecture so that they retained contact with nature in the form of terraced houses and the spatial element of *Bar-haiwan*. It was also found that the notion of social life in a Kurdish city is rather different from other Islamic cities due to the socio-cultural and environmental circumstances that have embraced this society and engendered a distinct notion of neighbourhood among its members, and thus, different attitudes to public life in respect to women's presence. Concerning the physical form of the city, these attitudes together with the features of the terrain gave rise to a soft boundary between the two realms of social life: the public and the private domains.

The circulation system of the neighbourhood revealed a relatively hierarchical order from the main focus of the city—the bazaar complex—to the principal route which supported the linear *Bazaarche*, then to the primary streets, *Kuche, Bunbast*, *Maidânche* and finally to the domain of private houses. This hierarchical order together with the patterns of women's presence in the public thoroughfares and the realms of *Maidanche* and *Kani* are summarised in the Fig. 5.8.

References

Abu-Ghazzeh, T. M. (1994). Built form and religion: Underlying structure of Jeddah Al-Qademah. *Traditional Dwellings and Settlements Review, V*(II): 49–55.

Abu-Ghazzeh, T. M. (1999). Housing layout, social interaction and the place of contact in Abu-nuseir, Jordan. *Journal of Environmental Psychology, 19*, 14–73.

Abu-Lughod, J. (1983). Contemporary relevance of Islamic urban principles. In A. Germen (Ed.), *Islamic architecture and urbanism* (pp. 64–70). King Faisal University.

Al-Kodmany, K. (2000). Women's visual privacy in traditional and modern neighbourhoods in Damascus. *Architecture and Planning Research, 17*(4), 283–303.

Alemi, M. (1991). Urban spaces as the scene for the ceremonies and pastimes of the Safavid court. *Environmental Design: Journal of the Islamic Environmental Design Research Centre, 11*, 98–107.

Allison, C. (1996). Old and new oral tradition in Badinan. In P. G. Kreyenbroek & C. Allison (Eds.), *Kurdish culture and identity* (pp. 29–47). Zed Books in association with The Centre of Near and Middle Eastern Studies.

Allison, C. (2001). *The Yezidi oral tradition in Iraqi Kurdistan.* Routledge.

Alten, I. (1958). *Sanandaj, a town plan.* United State Operation Mission to Iran.

Ayazi, B. (1992). *Ainah-yi Sanandaj*. Ayazi, Burhan.

Bahrami, B. (2001). *Tarahei Maskan dar Mahalla-ye Sartapula-ye Sanandaj* (Housing design in the neighbourhood of Sartapulah, Sanandaj). (Master), Art University of Isfahan, Isfahan.
Barth, F. (1953). Principles of social organization in Southern Kurdistan. Universitetets Etnografiske Museum. *Bulletin, 7*, 26.
Baskaya, A., & Symes, M. (1994). *A study of a house type with an open space. People, place and development*. CARDO.
Bruinessen, M. V. (1978). *Agha, Shaikh and state: On the social and political organization of Kurdistan*. Rijksuniversiteit.
Bruinessen, M. V. (2001). From Adela Khanum to Leyla Zana: Women as political leaders in Kurdish history. In S. Mojab (Ed.), *Women of a non-state nation: The Kurds* (pp. 95–112). Mazda Publishers.
Edwards, B., Sibley, M., Land, P., & Hakmi, M. (Eds.). (2006). *Courtyard housing: Past, present and future*. Taylor & Francis.
Galletti, M. (2001). Western images of women's role in Kurdish society. In S. Mojab (Ed.), *Women of a non-state nation: The Kurd* (pp. 209–225). Mazda Publishers Inc.
Grabolle-Çeliker, A. (2019). Kurdish women. In M. M. Gunter (Ed.), *Routledge handbook on the Kurds* (pp. 239–256). Routledge.
Guest, J. S. (1993). *Survival among the Kurds: A history of the Yezidis*. Routledge.
Gunter, M. M. (2018). *Historical dictionary of the Kurds* (Vol. 8). Rowman & Littlefield.
Hakim, B. S. (1986). *Arabic-Islamic cities: Building and planning principles*. KPI.
Hansen, H. H. (1960). *Daughters of Allah*. Purnell.
Haqshenas, M. (1999). *Sharakam Sennah* (Sanandaj; my land). Zhyar.
Hardi, A. (2004). The secret of solitude. *The International Journal of Kurdish Studies, 18*(1–2), 192.
Hassanpour, A. (1996). The creation of Kurdish media culture. In P. G. Kreyenbroek & C. Allison (Eds.), *Kurdish culture and identity* (pp. 45–84). Zed Books in Association with The Centre of Near and Middle Eastern Studies.
Hay, W. R. (1921). *Two years in Kurdistan: Experiences of a political officer 1918–1920*. Sidgwick & Jackson Ltd.
Izady, M. R. (1992). *A concise handbook: The Kurds*. Crane Russak (Taylor & Francis).
Katouzian, S. (1996). Tehran, capital city: 1786–1997 the re-invitation of a metropolis. *Journal of the Islamic Environmental Design Research Centre, 15*, 34–44 (In L. Micara & M. Alemi (Eds.), Como, Italy).

Kellett, P. (2003). Constructing informal places. In S. Menin (Ed.), *Constructing place: Mind and matter* (pp. 87–98). Routledge.
Khatib-Chahidi, J. (1981). Sexual prohibitions, shared space and fictive marriage in Shi'ite Iran. In S. Ardener (Ed.), *Women and space* (pp. 112–135). Croom Helm.
Kheirabadi, M. (1991). *Iranian cities: Formation and development*. University of Texas Press.
Kheirabadi, M. (2004). *Religions of the world Islam*. Chelsea House Publishers.
Kirmanj, S. (2014). Kurdish history textbooks: Building a nation-state within a nation-state. *The Middle East Journal, 68*(3), 367–384.
Kreyenbroek, P. G. (1996). Religion and religions in Kurdistan. In P. G. Kreyenbroek & C. Allison (Eds.), *Kurdish culture and identity* (pp. 85–110). Zed Books in association with The Centre of Near and Middle Eastern Studies.
Lawson, B. (2001). *The language of space*. Architectural Press.
Levkowitz, J. (2018). Palaces for the people: Scrutinizing social infrastructure in Sulaimani. *The Journal of Intersectionality, 2*(2), 24–32.
Lombard, M. (2014). Constructing ordinary places: Place-making in urban informal settlements in Mexico. *Progress in Planning, 94*, 1–53.
Madanipour, A. (1998). *Tehran: The making of a metropolis*. Wiley.
Madanipour, A. R. (2003). *Public and private spaces of the city*. Routledge.
Mazumdar, S., & Mazumdar, S. (2001). Rethinking public and private space: Religion and women in Muslim society. *Journal of Architectural and Planning Research, 18*(4), 302–324.
Memarian, G., & Brown, F. E. (2003). Climate culture and religion: Aspects of the traditional courtyard house in Iran. *Architecture and Planning Research, 20*(3), 180–198.
Mirawdeli, K. (2003). Claudius Rich's Kurdistan in 1820. *Kurdistanica* (The Encyclopaedia of Kurdistan), History: Related articles. Nottingham. https://www.kurdistanica.com/index.html.
Mojab, S. (2003). Kurdish women. In S. Joseph (Ed.), *Encyclopedia of women and Islamic cultures: Family, law and politics* (Vol. 2, pp. 358–366). Brill.
Monteguin, F.-A. D. (1983). The essence of urban existence in the world of Islam. In A. Germen (Ed.), *Islamic architecture and urbanism* (pp. 43–63). King Faisal University.
Morris, A. E. J. (1994). *History of urban form: Before the industrial revolutions*. Longman Scientific & Technical and Wiley.
Moughtin, C. (2003). *Urban design: Street and square*. Architectural Press.

Nikitine, B. (1956). *Les Kurdes. Etude sociologique et historique*. Imprimerie nationale, Bibliothèque C. Klincksieck.
Rainer, R. (1977). *Traditional building in Iran*. Akademische Druck- u.
Rangavary. (1999). City's memories/Interviewer: M. Haghshenas, S. Marufi, & S. Munaghi. (Vol 1). Sanandaj.
Rich, C. J. (1836). *Narrative of a residence in Koordistan, and on the site of ancient Nineveh: With journal of a voyage down the Tigris to Bagdad and an account of a visit to Shirauz and Persepolis*. J. Duncan.
Saleh, M. A. E. (2001). Planning and designing for defense, security and saftey in Saudi Arabian residential environments. *Architecture and Planning Research, 18*(1), 39–58.
Southall, A. (1998). *The city in time and space*. Cambridge University Press.
Stewart, D. J. (2001). Middle east urban studies: Identity and meaning. *Urban Geography, 22*(2), 175–181.
Sultanzadeh, H. (1986). *Moghadamai Bar Tariekhi Shahr va Shahrnishini dar Iran* (An introduction into the history of the city and urbanisation in Iran). Amir Kabir Press.
Sultanzadeh, H. (1994). *Entry spaces in traditional architecture of Iran*. Municipality of Tehran.
Tallan, C., & Taitz, E. (2009). *Learned women in traditional Jewish society in Jewish women: A comprehensive historical encyclopedia*. https://jwa.org/encyclopedia/article/learned-women-in-traditional-jewish-society. Online: Jewish women's Archive.
Tavassoli, M., & Bonyadi, N. (1992). *Urban space design 2*. Urban Planning and Architectural Research Centre of Iran.
Vasilyeva, E. I. (2000) The Kurdish city of Sanandaj: Ardelan's Capital. Paris, conference presentation.
Warren, J., & Fethi, I. (1982). *Traditional houses in Baghdad*. Coach.

6

Private Domain

Abstract This chapter aims to investigate the private domain to grasp the continuity of the order perceived in public spaces into the private spaces, particularly to the courtyard part of the houses. In this part, the means of connection to the outside world and two types of houses' spatial organisation including higher and lower status families have been analysed because this made clear different values and attitudes towards the place construction. This comparison reveals some points in connection with the nature of Islamic cities in Iranian context, because the houses of notables can be interpreted in line with the general model of Iranian cities with introverted structure due to their socio-political relationships with other Iranian cities based on their status, while the commoner's dwellings are more indigenous with the spatial features of extroverted structure.

Keywords Private domain · *Bar-haiwan* · Indigenous architecture · Houses · Courtyard

H. Alizadeh, *Building a traditional Kurdish City*,
https://doi.org/10.1007/978-981-16-3634-9_6

Introduction

Considering this general notion that the external order is meaningless without regarding the internal order because "the way the external order is developed depends on what happens in internal spaces and vice versa" (Rapoport, 1977: 309), this chapter aims to investigate the private domain to grasp the continuity of the order perceived in public spaces into the private spaces, particularly to the courtyard part of the houses. In this part, the means of connection to the outside world and the spatial organisation of the houses into two realms of higher and lower status will be analysed because this makes clear different values and attitudes towards the place construction. This comparison will also reveal some points in connection with the nature of Islamic cities in Iranian context, because the houses of notables can be interpreted in line with the general model of Iranian cities due to their socio-political relationships with other Iranian cities based on their status, while the commoner's dwellings are more indigenous.

As aforementioned, the transitional connection space to the private domain in the circulation system of the neighbourhood revealed different structures related to the types of houses. In the houses with an introverted structure, belonging to the nobles and rich people, another space was maintained immediately before the doorway to the house. This space is rather like a *Nema-hashti* (a semi-polygonal uncovered space) which acts as a transitional space between the two realms (Fig. 6.3). It is also a shared space with the element of *Khaje-neshen* or platform. In the houses of commoners, the size of *Nema-hashti* was either minimised or omitted from its structure depending on the wealth of the owner. The details of these features in two groups of houses are reviewed in the following sections.

Houses of the Ruling Classes and Notables

In the first group, there is a hierarchical spatial order to approach the main courtyard of the house somewhat similar to the form of house in the Iranian central plateau (Fig. 6.1). This pattern reinforces the concept

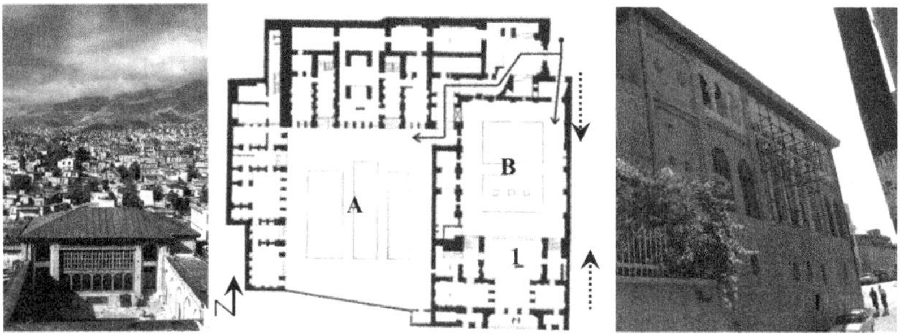

Fig. 6.1 Different pattern of spatial organisation in the houses of ruling class (*Sources* Plan and left photo from Gazarani collection and right photo by author autumn 2004)

of privacy as it restricts direct relationship between the interior and the exterior of the houses. This way of connecting the exterior to the interior of the houses was recognised as one of the main architectural principles in Islamic belief (Moradi & Amirkabirian, 2001). In this pattern, the houses' privacy was usually maintained by a *Hashti* (vestibule) immediately after the entrance door, as a stopping point to give directions or as "a temporary reception room for those persons who did not need enter the house" (Memarian & Brown, 2003: 189). This is followed by an angled corridor linked to the other parts of the house, especially to the corner of the courtyard. Depending on the status of the owner, this corridor has also a link to the stable. In some cases, due to the socio-political status of the owner, this spatial relation was complicated by the creation of a broad division of the house into two quarters of men and women—thus creating two courtyards. The spatial structure defines the realm of houses in this category into two separate parts termed as the *Birun* (outside/public) and the *Andarun* (inside/private) in respect of sex segregation (Mazumdar & Mazumdar, 2001; Memarian & Brown, 2003).

In this category, there are other houses within which this spatial relationship had shrunk into a simple form just to prevent a direct view into the house by offsetting the corridor or creating another barrier, a second door, at a distance behind the entrance door.

The spatial organisation of the house's components was mainly affected by the physical conditions, both climate and topography. In Sanandaj, the climate condition was recognised as a semi-arid one so that its summer usually has relatively moderate conditions but the winter is very cold (Far-Afza, 1991). The harsh weather of winter, with freezing days, usually lasts for 100–120 days from mid November until the end of February. Hence, the main parts of the house—the living quarters—were positioned on the north and west sides of the courtyard, where they are most exposed to the sun during the winter. For this reason, the orientation of the courtyard was directed towards the south and in some cases towards the east. In line with this orientation, the main part of the house, which contained the element of *Bar-haiwan/talar*, was aligned predominantly on north or west sides of the courtyard and the other sides maintained for the service rooms with lesser spatial dimensions.

In spite of the above point that characterises the orientation of the houses in response to the climate and climatic gradients, there are some other cases where the main part of the house was moved to the south side because of topography. This granted such houses a possibility of viewing the natural scenery as they maintained an opening with a view towards the Awear Mountains (Figs. 5.6 and 6.1). In these two photographs and the map in Fig. 6.1, the means of access to the thoroughfare and to maintaining a view of the scenery had important roles in the location of the main parts of the houses and the form of spaces. In the first house (A, Salar Saeed's house) the main part was positioned on the north side of the courtyard and its south side was kept blank to retain the view of the Awear Mountains to the south. But, this was changed in the second house (B, Sheikhul-Islam's house) as it has a direct relation to the thoroughfare. Although, the main part was aligned with the south side of the courtyard, the view of scenery was maintained by an elaborated Bar-haiwan (right photo) looking towards the distant sceneries of the Awear Mountains linked to the sash-windowed room overlooking the courtyard (1, Otaghi Orsi).

Overall, this group of houses can be recognised as an expression of formal architecture within which the spatial distribution of the spaces is consistent with the concept of privacy and the status of the owner. The courtyard structure, with the pattern of four-sided exclusion from the outside world, was reinforced, thus limiting the level of fluidity between two realms of urban life. In this kind of architecture, it is difficult to observe the socio-cultural relationships between neighbours within the house and even in the spaces linked to the outside, the *Kuche*, as the relationships are more formal and depend on the status of the occupier. And, more importantly, the space organisation does not lend itself to ease relationships between neighbours. As aforementioned in the concept of *Zheer-mala*, the only place for women to gather, particularly in the houses of rich people but not those in charge of governmental affairs (ruling class), was the place of *Sar-tanour* (the place of the oven) close to the entrance door. It is noteworthy that the large *Nema-hashti* (a semi-polygonal uncovered space) and the platforms located within the entry spaces of these houses were not pleasant for socialising among neighbours. They mostly functioned as places of rest for visitors because they had to wait for a while to receive permission to enter or to conduct affairs.[1] Apart from this function, due to the status of the ruling-class families, who possessed properties in the countryside and, thus, these entry spaces were usually used by the *Khan* and his peasants or servants. While he sat on one of the benches/platforms—the main one close to the doorway, those people were there to attend to his orders.

Houses of Commoners

In the second group of the houses, as explained earlier, the hierarchical spatial movement was reduced to its simplest form in order to ease communication among neighbours. As Memarian & Brown (2003) indicate, this is in relation to informal relationships and more importantly to the intimacy and hospitality required by Iranian culture to

[1] In line with this point, one can look at the entrance doors furnished with two splendid knockers in this group of houses: one with a metal ring for women, the other a large metal hammer for men. For Marufi (Interviewed 29.08.2003) this kind of sex segregation pattern is not from Kurdish culture. As he emphasised, it is more an imitation from other Iranian cities because this high level of sex segregation was not of such importance for Kurdish people. The evidence can be traced through the settlements in rural areas of the region which lack such separated knockers.

visitors, especially among the commoners. In line with previous discussions concerning the strong sense of neighbourhood among Kurdish people and the inhabitants of the *Qatar-chyan*, this attitude to intimacy and hospitality can be recognised as a remarkable character of Kurdish culture. For example, as Barth (1953: 109) indicates, "hospitality is an integral aspect of the Kurdish ideal of a man, conversely, a person with a reputation for miserliness suffers corresponding loss of prestige". This is what impressed Rich (1836: 70) in his first contact with the Kurds in 1820. For him, "the manners and customs of the Kurds [are] unlooked-for honour, and [are] great proof of his friendly and hospitable disposition".

In spatial terms, the organisation of the spaces is not in the rigid and formal structure similar to the first type of the houses so that in many cases the courtyard directly opens onto the thoroughfares or to the Maidanche at the end of the Bunbast (Fig. 5.8). This can be related to different factors, to the size of the plots, to the status of the owner, to the location of the house and to the sense of neighbourhood among the residents. The location of the main parts of the house depended on the location of the house in relation to the thoroughfares and the slope of the site. In the relatively level part of the neighbourhood, the north and west sides of the courtyard contained the main part of the house. Although, this pattern can be recognised as logical in responding to the climate, it changes as one approaches the sloping part facing the north. In this part, the factor of slope had a major role in the orientation of the houses as the south part was used to accommodate the main parts, especially the prominent spatial feature of the Bar-haiwan (Fig. 5.6, bottom left photo). Considering this point, one may ask that how this kind of house was adopted with the climate conditions, especially the harsh winter weather. Based on the master plan of the city, the prevailing direction of the wind is towards the north. This to some extent moderated the harsh conditions of the weather in the winter. In addition, because of this weather, the two-storyed houses were generally built cubically, with thick walls and doubled ceilings as the best solution for controlling the internal heat and to provide bearable conditions to the inhabitants during winter.

In contrast to the first group and what Memarian and Brown (2003) indicate in relation to the form of houses in Yazd and Shiraz, they are more intimate to the public world, especially to the close neighbours as they had two stories with the possibility of a direct and outward-looking contact with the world of the kuche (public) (Figs. 6.3 and 5.6, bottom left photo). In line with the above points and as a result of the fieldwork studies and interviewee's sayings, there was a custom among the close neighbours to leave open the entrance doors during the day for the sake of offering a welcome to any guests[2] and of easy communications among the neighbours in Qatar-chyan.

> There was a strong trust among us. For this reason, our interaction was easy. Every day, it was a duty for us to know what is going on for our neighbours. I remember the period when my father was in trade with the city of Hamadan. It usually took a long time, nearly one month, for him to return. For that period, one of our neighbours, called Housein, came to visit us every day in the early morning in order to be sure about everything and what I needed for that day. (Khalil Shahla[3])

By comparing the plans of those two types, it is possible to derive the basic plan of the houses which usually comprises a tripartite structure, called *Se-bakhshi* by the people. The basic spatial structure of this plan comprises a *Bar-haiwan* in front of the main living room flanked by two rooms (Fig. 6.2). The *Bar-haiwan* usually gives access to other living spaces. This three-plan house can also be found elsewhere in Iran. It is more similar to the plan that Warren and Fethi (1982: 30) derived from the study of traditional houses in *Baghdad*. The point of difference is in relation to how the plans are related to the realm of the public domain in respect of the concept of privacy. The case of *Baghdad* was characterised by balconied windows which jutted out from upper rooms while having a special veil meant as a barrier called "*Shanashil*"/an elaborate

---

[2] The Kurdish people are usually well-known for their hospitality among the other Iranian people. This together with the location of the *Qatar-chyan* neighbourhood, which was separated from the main city by the *Dara-bayan* River, gave rise to a strong solidarity among its inhabitants to such an extent that the majority of people are still conceiving their past life by referring to the way of life of this neighbourhood.

[3] Interviewed 14.09.2004, he was around 80 years old and one of the merchants, who previously lived in the neighbourhood of the Bazaar with his family. As he explained, his family is originally from the Iraqi part of Kurdistan and migrated to the city of *Sanandaj* during the *Qajar* dynasty.

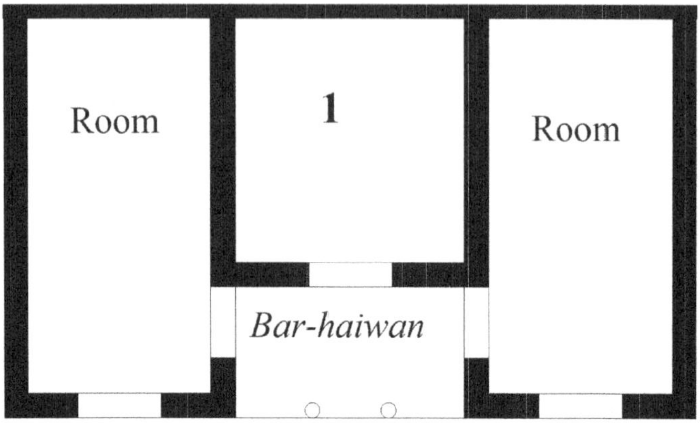

Fig. 6.2 The basic plan of house design. Depending on the status of the owner, the main part (1) can be changed. In higher status houses, it was normally used as the main part of the house (guest room) consisting of a living room or sash-window room (Otaghi orsi) overlooking the courtyard or bar-haiwan in front. In the lower status homes, it is either a main living room or it was squeezed into a corridor (usually with a closet at the back) to link two surrounding rooms (*Source* author, Autumn 2005)

overhanging screen, which prevented any view from outside to inside while the sun could be allowed in and the inhabitants could see the outside. On the contrary, as explained, the case of *Sanandaj* is free of this screen in the realm of public domain and instead of a jutting out pattern it was set back to define a transitional space, as a softer barrier and welcoming space, to negotiate with the outside world.

The origin of the basic plan above can be traced in the farmhouses of the countryside (as described in the previous chapter with respect to the second context of Kurdish social life), especially in the villages to the south and south-west of the city, where the slope gave direction and structure so that the roof of one house acts as the courtyard of other house and the throughway of the settlements, thus it was appropriate for them to keep a broad view of distant scenery. Here the spatial structure of the houses is such that it eases communications between the neighbours. It means that the house is in direct connection with the *Kuche* (Fig. 6.3) which invites the neighbours to sit either in front of the house (*Bar-mal*) or in the main part of the house specified as being mainly for

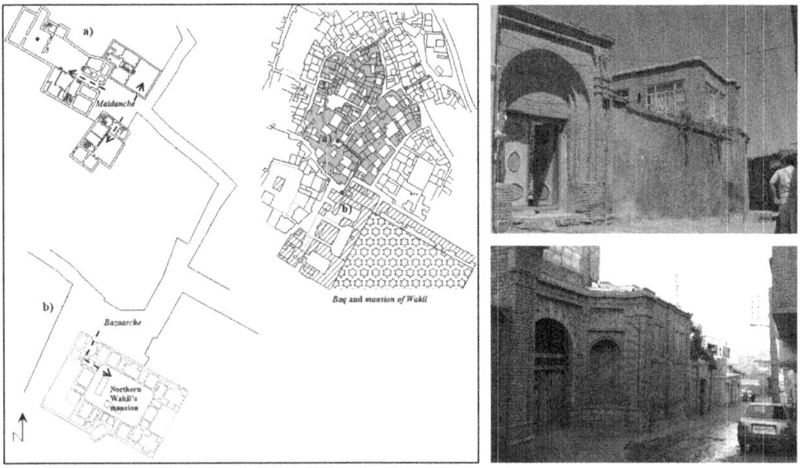

Fig. 6.3 Summary map concerning the points of differences in two groups of the houses and their connection to the realm of Kuche (*Source* Author, map Autumn 2005, photos August 2004 and September 2018)

the guests (Figs. 5.4 and 6.3). This kind of house aggregation and its neighbouring structure implies the previous concept of rooftop society common in the social life of the Kurdish people, fluidity between inside and outside worlds. Due to such a socio-spatial background and the fact that the indigenous houses mostly contained a small floor area, the *Housh* (courtyard) was not adequate to support the social life of the neighbours, thus, the *Bar-mal* was the replacement to entertain most of the neighbours and to maintain contact with outside world, and possibly with the distant scenery. Alongside the *Bar-mal*, and mainly to do with women's socialising, the house was the arena for regular visits among the neighbours, especially during the winter nights in the main part of the house, the guest room, and on summer nights on the rooftop. There, while the people were entertained with tea, cigarettes and dried fruits, usually by the women, the elders reminisced, and there was a kind of storytelling and the young would listen carefully (Ayazi, 1992). Sometimes too, the atmosphere would change somewhat and all men present, young and old, would play communal games called *Gourwa-bazee*—guess who has the ring.

Conclusion

From the above review concerning the two groups of the houses and their comparison with the notion of the *Maskan*[4] in traditional courtyard houses of other Iranian cities, it can generally be argued that the spatial structure of the first group was more in line with the concept of privacy. This restricts the ease of communications between the inside and outside worlds. This is more due to the socio-economic relationships of the notables and rich people with other Iranian cities which contextualised such pattern of house configuration in Sanandaj. More importantly, the role of those people migrated to the city, especially those invited by the *Wali*, is the most significant. The convincing evidence for this is a statement in history which says;

> *Amanullah-khan* the great (*Wali* of *Sanandaj* in 1800–1825) had been enamoured with the architecture of Isfahan regarding his visits to this city. For this reason, when he met an *Isfahanian* merchant (*Abutaleb Isfahani*) in the city of Tehran in 1808, he invited him to settle in the city of *Sennah* and help him to renovate the city based on the quality of architecture in the city of Isfahan. *Abutaleb* settled in the city the year after and then he invited architects, carpenters, tilers and others to *Sanandaj*.[5]

On the contrary, the basic pattern of spatial structure and means of orientation to the outside world in the second group are in line with Kurdish culture was borrowed from the indigenous architecture of the region, which implies a kind of fluidity between two realms regarding the effects caused by the earth and the sense of neighbourhood.

---

[4] *Maskan* derives from the root *Sikun* which literally means Quiet (Warren & Fethi, 1982: 44).

[5] This was quoted in a report concerning the Mansion of *Asaf-Vazieri* in the library of KCHO. Although the date and author of this report is unknown, the point mentioned, especially the construction of the city's main edifices by Isfahanian's architects, was also mentioned in the historical books such as Mardukh (2000), Kurdistani (2002) and Rich (1836).

References

Ayazi, B. (1992). *Ainah-yi Sanandaj*. Ayazi, Burhan.

Barth, F. (1953). Principles of social organization in Southern Kurdistan. Universitetets Etnografiske Museum. *Bulletin, 7*, 26.

Far-Afza. (1991). *Tarh-e Tawseah va Omran va Hawzaieh Nufuz va Tafseli-e Shar-e Sanandaj* (Sanandaj's comprehensive and master plan). Sazman-e Maskan va Shahrsazi.

Kurdistani, A. V. (2002). *Hadigha-E-Nasriah dar Guqkrafia VA Tarikh-e Kurdistan* (History and geography of Kurdistan) (Persian). Tavakouli.

Mardukh, M. (2000). *Tarikh-e- Mardukh* (The history of Kurds and Kurdistan). Karang.

Mazumdar, S., & Mazumdar, S. (2001). Rethinking public and private space: Religion and women in Muslim society. *Journal of Architectural and Planning Research, 18*(4), 302–324.

Memarian, G., & Brown, F. E. (2003). Climate culture and religion: Aspects of the traditional courtyard house in Iran. *Architecture and Planning Research, 20*(3), 180–198.

Moradi, A. M., & Amirkabirian, A. (2001). *Muarafei Teadadi az Abniea-e Sunat-e Iran va Tahlili bar Vizhagihaie Fazaei Anha* (Introducing a number of traditional buildings in Iranian architecture and analysis of their spatial characteristics). Pazhuheshgah.

Rapoport, A. (1977). *Human aspects of urban form: Towards a man-environment approach to urban form and design*. Pergamon Press.

Rich, C. J. (1836). *Narrative of a residence in Koordistan, and on the site of ancient Nineveh: With journal of a voyage down the Tigris to Bagdad and an account of a visit to Shirauz and Persepolis*. J. Duncan.

Warren, J., & Fethi, I. (1982). *Traditional houses in Baghdad*. Coach.

7

Conclusion

Abstract As each chapter contained a conclusion section, this chapter draws attention to the important points summarised from the whole book. It outlines the findings with some recommendations for future studies. It also draws attention to some suggestions concerning the findings and their application in future prospects of the city and similar situations. Within the wider context, a comparative approach within Kurdish cities can highlight similarities and differences to form their main socio-spatial characteristics and underlying design principles as the components bases to define the concept of Kurdish city in general.

Keyword Kurdish city · Design principles · Kurdish culture · Experiential qualities of space · self-reliant character · administrative structure

The main concern of this book was to understand the nature of Kurdish settlements by paying attention to the interaction between the multifarious social and spatial forces that have conditioned the processes and patterns of Sanandaj's formation and development in the course of time.

H. Alizadeh, *Building a traditional Kurdish City*,
https://doi.org/10.1007/978-981-16-3634-9_7

It provided a comprehensive morphological study of Sanandaj to respond to the clear need to consolidate the scarce formal literature and local documentation on the Kurdish cities in general and Sanandaj/Senna in particular. With the above points in mind and the fact that each chapter contained a conclusion section, this chapter is organised to draw attention to the important points summarised from the whole book.

Keeping in mind that "the social premises of designers" (Kostof, 1991: 11) are important in the analysis of traditional built form formation, the historical background outlined the long process of urban evolution. It highlighted the changes in the built fabric which have produced a certain richness and uniqueness of historic meaning and importance associated with the built form of *Sanandaj*. It became clear that the emergence of the city was more in relation to the conflicts that happened between the two rival empires on either side of the *Zagros Mountains* so that *Sanandaj*, as the capital of the *Ardalan* family, was founded in 1639 as a politico-administrative centre supporting *Safavid* policy. Also, the concept of mound cities was conceptualised not only as a background to the city's emergence, but also as the leading structure borrowed from a pre-Islamic concept evident at the time of the Median dynasty (728–550 BCE), and, even before the time of this dynasty appearing as a typical style of urban planning among the civilised people living in the *Zagros-Taurus* mountains ranges (Izady, 1992). The recalling of this concept, as a leading structure of Kurdish cities, led to the emergence of a unique structure in the foundation of the city that no other Iranian cities possessed in the periods of the *Safavid* (1502–1736) and the *Qajâr* (1794–1925) dynasties.

Within the second chapter, first the details of place structure in both man-made and natural settings were investigated to make clear the socio-spatial determinants of the built form in general. Within this section, first the geophysical setting of the city was analysed through analytical drawings to identify the structuring elements, those features of the natural setting which have influenced the city's orientation and its built form in general. Two levels of edgedness were characterised as the macro and micro geophysical settings which defined the city's boundaries analogous to the multiple walls of the city. This part reveals that the nature, as "the primary components of the given" (Norberg-Schulz 1980: 10), constrained the city's main orientation towards the east where the sacred hill of *Toos-nauzar* is located and made this orientation stronger than usual so much so that most of the main public buildings (bazaar, *Jamia* mosque, *Qala*/citadel, garden and mansion *of Khusrau-awa,* and *Wakils'* mansion) were oriented towards the east in general and this hill in particular.

Following the geophysical setting, the morphological components of the city were analysed, to obtain a better understanding of the city's urban form. Based on the notion of Kostof (1992), the connection between the constituent elements of urban form and that of the natural setting was examined. It became clear that the city's walled pattern, which only included the central part—the city of nobles/Khans—was recognised as a unique pattern in Iranian cities, which stemmed mainly from the concept of mound cities.[1] This form of city's structure, which dominated over the territory of Ardalan rulers in the region of Kurdistan for nearly five centuries before the city's foundation in 1639, was recognised as a background to the city formation. Due to such structure which strengthened the mound cities security, the region of Kurdistan enjoyed an independent status on the "Perso-Turkish" frontiers (Minorsky, 1943). That is why the founders of the city, the *Ardalan's* family, regenerated such structure to improve their security again in the new founded city of Sanandaj. Based on this structure, the city's general spatial organisation included three main parts: (1) the *Qala*/citadel with its own embattled walls centrally located on the top of a hill. (2) The the inner-city (*rulling-class quarter*) in the central part of the whole city, again with enclosing wall. (3) The outer city (*other neighbourhoods*), which formed the third layer of the city's fortification in general. The central part of the city was conceptualised as "a real city inside the city" because it contained the ruling class' houses with their own urban facilities and was separated by a fortified wall—the second wall—from the city of the commoners. Its connection was maintained through four fortified gates set at the cardinal points of the compass with some degrees of deviation set in the direction of the city's neighbourhoods, which can be recognized as the symbols of

[1] The structure of mound cities was characterised as a leading structure borrowed from the ancient style of urban planning among the Kurdish people within the *Zagros-Taurus* mountain ranges (Izady, 1992).

Khâns' authority. Considering such a structure, especially the seclusion of inner-city from the commoners, one may propose the notion of class division in a way similar to the structure of *Sassanid* cities (224–642 CE). In response to this point, it is worth to say that "belonging to the quarter and living in it could and must be associated with a certain social status and even with a political orientation" (Vasilyeva, 2000: 8). This means that the stratification of notables (*Khans*) and common people was not based on wealth alone because a number of persons among the commoners could be very rich.

The outer city was conceived as falling into four neighbourhoods. Each neighbourhood was organised on the basis of specific social status and professional background of the inhabitants, and they circled the nucleus of power in the form of satellite settlements. This pattern of organisation of the neighbourhoods distinguished the city from other Islamic and Iranian cities where the division of neighbourhood is mainly based on family ties, clientage, common village origin, and ethnic or religious sect identity.

Such organisation of the city's internal divisions, which to some extent reflected the inhabitants' social status, was perceived as being consistent with some formal planning in the foundation of the city because, as Akbar (cited in Morris, 1994: 381) indicates, "the term planned applies only in respect of a political decision to establish a permanent settlement on a given site". Accordingly, the ruling houses—the controllers of the political fate of the region—occupied the city's core—the inner-city. The neighbourhood of bazaar, which carried the economic fate of the region, was organised at the bottom of the *Qala*'s hill to its east. The muleteers were accommodated on the other side of the river to the south and detached from the rest of the city due to their function. This was justified in relation to the sanitary reasons and the presence of an opposition group with a political orientation against the ruling class. Finally, the servants due to their low status were pushed onto the hilly parts of the city's geographical setting in the north, where they had a direct connection to the principal road into the inner-city. Due to such a diversity of social background among the neighbourhoods, the urban form of each one was also regarded as communicating its own character reflecting the social status of the inhabitants. For instance, the street pattern in the neighbourhood of the bazaar was shaped in a maze pattern analogous to the beehive's cells, which contained a small number of *Bunbast* (cul-de-sacs). This pattern afforded more security to the merchants and minorities as its circulation system disoriented strangers and hindered invaders. However, the street pattern in the neighbourhood of *Jawr-awa* was mostly oriented by topographical features of the site and was free of any notion of defence because of the low status of the inhabitants. Generally speaking, the organisation of place described above shows that both rational and organic patterns were represented in the city of *Sanandaj*.

For the city's public places, three *Maidans* were identified comprising *Maidan-e Arg* in front of the *Qala*, *Maidan-e Alaf-Khan* formed by the rectangular form of the Bazaar, *Maidan-e Kilisa* (Church Square) in front of the church in the neighbourhood of *Aqa-zaman*/bazaar. The results of the analyses indicate that the *Maidan-e Arg* and its function were to some extent similar to the squares in other Iranian cities, especially those cities with significant strategic importance in their region. *Maidan-eAlaf-Khan* and the bazaar were characterised as a unified urban element because the form of the bazaar created such a *Maidan*. Although the root of such a concentrated form of bazaar was traced in other Iranian cities and related to the *Mayadin-e Naqsh-e Jahan* in *Isfahan* and the *Ganj-ali-khan* in *Kerman*, it was recognised as a unique concentrated bazaar in the Iranian context because the lines of bazaar were just maintained around the four sides of the *Maidan* without any lines towards the *Maidan-e Arg*. The reason for such a design was ascribed to the hilly features of the city's geophysical setting which rather limited bazaar's linear form and the political dimension of place-making which outweighed its economic dimension. For such reasons, unlike other Iranian cities, the bazaar was not formed along the principal road, the city's backbone, which was led from north to south. Apart from the topographical features of the terrain, the reason for such a structure was defined in terms of the city's defense pattern. As aforementioned, the third part of the city lacked a man-made fortified wall, thus it sandwiched the bazaar by its compact form without a direct arterial road towards suburbs.

Following the city's public places, the principal road of the city was identified. It was the backbone of the city in structuring the neighbourhoods (*Jawr-awa, Char-baq, Meyan-qala* and *Qatar-chyan*) and various other physical and spatial urban elements. It ran from north to south crossing the *Dara-bayan* River towards Awear Mountains. The *Dara-bayan* River, as the structure of natural setting, defined the longitudinal axis and the principal road, as the man-made environment, structured the latitudinal axis. These two linear forms had major roles in the organisation and orientation of the city's spatial structure. Contrary to the common notion that the course of principal road was mainly bounded with the structure of bazaar, here in Sanandj, the principal road oriented by the poles of powers within the inner-city based on the Qala and suburb on the slope of the *Awear Mountains* where the second seat of power was located (the mansion and garden complex of *Khusrau-awa*) along which the *Arg* square and the *Qatar-chyan bazaarche* evolved.

The last public urban element discussed was the *Jamia* mosque. It was an important element of the city to such an extent that its history defined the life of the city as well. Its location was changed over time from the arena of the bazaar to the inner-city close to the site of power due to its political dimension. The rulers were aware of a constitutional role of *Ulama* and *Bazaaris* in social uprising against any unacceptable decisions, therefore, the *Jamia* mosque was detached from the city of commoners to enable them to control any social unrest. This made it also unique in the Islamic and Iranian cities, where the design principles required Jamia mosque to be adjacent to the site of the bazaar.

For details of place structure, the *Qatar-chyan's* morphological analysis highlighted two concepts of point and line. The existing passway to *Awear Mountains* was intensified in time by another pole of gravity as a single point in space—the Mansion and garden complex of *Khusrau-awa*—on the slope of these mountains. As a continuation part of the principal road, it also formed the *Qatar-chyan* linear *Bazaarche* and attracted more people to settle in its environs. This means that the creation of the line was mainly bound up with the power relation focused on the *Wali*. Other social forces were identified in the neighbourhood evolution including two great families. The presence of *Wakil* family—one of the notable families was driven out of the inner-city—affected the neighbourhood's development towards the south and their mansion become a focal point for the second centre of *Qatar-chyan* in the form of a concentrated centre. The *Galadari* family—as cattlemen and muleteers whose business was in relation to the function of the bazaar—settled in the west part of the neighbourhood and formed another point of gravity alongside other forces of development. The *Dara-bayan* River, which divided the city into two separated parts and bordered the northern part of the neighbourhood, was conceptualised as an important feature of the terrain effected the orientation of the houses towards the north and as they also maintained the spatial element of *Bar-haiwan*. This was recognised as a striking feature of the settlements in the *Zagros Mountain* Range, which was embedded in Kurdish culture and, thus, transferred to the urban settlements of the region.

For the social dimension of place, two public features were characterised as they influenced the social life of Kurdish people as well: religious and secular elements. The first element was, in a sense, analogous to other Islamic cities where adults and young men gathered in the mosques for daily prayers. This manner of gathering institutionalised regular communication which strengthened the sense of neighbourhood, mutual responsibility and social cohesion among the inhabitants. This, together with the presence of leading *Ulama,* led to a substantial degree of internal self-government without recourse to the external regulatory agents in the neighbourhood. The second element, the *Mewan-khana*, was identified as a central institutional element in the evolution and transmission of Kurdish habits and traditions;

> Information reaching the community would be discussed in such gatherings, [Mewan-khana], and to some extent assimilated to the communal culture. (Kreyenbroek, 1996: 105)

It is a social element which conveys the hospitable character of Kurdish society. In *Qatar-chyan*, the traces of this element were identified in the character of *Chaikhana/Qahve-khane* where, similar to the *Mewan-khana*, the adults and young men spent their leisure time drinking tea, talking about the events of the day or listening to the narrative of a storyteller and, thus, the young were taught traditions and etiquette. The importance of this social element was such that two centres of the neighbourhood were predominantly characterised by the presence of the element of *Chaikhana* alongside the religious elements of the mosque and the *Hammam*.

Following the men's social realm, the women's social realm was explored, first by examining the notion of privacy in Kurdish culture. The notion of privacy was mitigated by the nature of Kurdish culture which had evolved within the *Zagros Mountains* where the geophysical setting and the rich colours of the landscape with the long unstable history of Kurdish people gave rise to distinctive traditions and habits (Izady, 1992) namely a strong sense of neighbourhood and of hospitality. As a result, the tension between public and private domains was somewhat reduced so that the women could be present and interact in the public spaces of the neighbourhood. The primary relationships occurred within the neighbourhood "where there [was] very little privacy and anonymity" (Rapoport, 1977: 297) due to the homogeneity of the inhabitants and strong sense of neighbourhood among the neighbours. The effects of such attitudes to privacy were considered with respect to the nature of interaction between the two realms so that the indigenous houses characterised as extroverted structures intimate with the world of the public spaces by means of the spatial element of the Bar-haiwan and openings overlooking the public domain and even a neighbour's houses, and courtyards open directly to the Kucheha/alleys. This means that the harmful notion of visual intrusion, which was emphasised in other Islamic cities, was not so important in the Kurdish urban context.

The widely held notion concerning the restrictions on women's presence in the public domain of Islamic cities was examined in the context of the neighbourhood. It become clear that this notion was more in line with the life of the rich and notables, while the commoners, who comprised the greater part of the city population, attended and interacted in the public spaces regarding the nature of their life (Alizadeh, 2007). Of the gathering public places, two places of the *Kani/spring* and the *Bar-mal/doorstep*—were somewhat peculiar to the Kurdish context and were the locations of the greatest part of women's social interaction. *Bar-mal* was identified as a desirable attitude for commoners due to their strong sense of neighbourhood and social cohesion and the spatial pattern of their settlements. On the contrary, it was disregarded by wealthy and *Ulama* families due to their status, values and lifestyles and the spatial pattern of their houses.

The street pattern of the neighbourhood shows a hierarchical system branched out from the principal road, the hub of the social interactions of the neighbourhood, to the primary and secondary *Kuchehas*, and finally to the *Bunbast*/cul-de-sac or the *Maindcnche*. These two last elements provided access to the cluster houses. The rationality of urban form was conceptualised by referring to the richness of contrasting situations, not to the monotonous geometrical forms. This was due to the tension between geometric and organic forms defined as binarism in the evolution of the city, particularly in the formation of line and point conceptualised in relation to the principal road and two poles of powers. The experiential qualities of space in the circulation system of the *Qatar-chyan* revealed two important lessons concerning the duality of space was rooted in the social structure of the neighbourhood, poor and rich living alongside each other, and the characteristics of the terrain. The last aspect was conceptualised as the synthetic order of place formed by the overlapping of places or the terraced pattern of the houses. The duality of space was mainly due to the existence of two attitudes to privacy among the inhabitants of *Qatar-chyan*. This was evident in the different features of the built form beside the public spaces—thoroughfares. The houses of rich and notables' families embraced an introverted structure similar to the concept of the courtyard pattern in other Iranian cities. But for commoners, an extroverted structure was a popular spatial pattern for their houses borrowed from the indigenous architecture of the settlements in the region. In this type of house, the spatial feature of *Bar-haiwsn* in contrast to its counterpart of the *Eyvan* in other Iranian cities is an important transitional space between the inside and the outside world. This means that the spatial element of *Bar-haiwan* served to keep contact with the natural landscape rather than the courtyard. Apart from those features mentioned, the way entrance doors were located along the streets shows a particular pattern in which the entrances were exposed to the view of passers-by. This enriched the visual qualities of public thoroughfares by differing images of the entry spaces—the last spatial element approaching the private domain—with a diversity of scale and portal decoration regarding the social structure of the neighbourhood where rich and poor lived side by side each other.

The last part of the book dealt with the private domain within the cluster houses as the primary setting of the family life. The analysis focused on the form of connection between two realms of society based on perceived socio-spatial characteristics. Generally speaking, two patterns of connection were identified: those with an introverted spatial structure recognised in line with the life of higher status families and analogous to the spatial structure of houses in other Iranian cities; and those with an extroverted structure characterised in line with the life of commoners. The first group of houses was usually typified by a large entry space beside the public thoroughfare referred to as the *Nema-hashti*. This is a transitional spatial element between two realms linking to the *Haush*/courtyard through a complicated hierarchy of spatial elements. In contrast, the second group was characterised by the simplest form of connection to the public realm so that the *Haush* directly linked to the *Kuche*/*allye* and contextualised easy communications between the neighbours. This was explained in terms of the nature of Kurdish society in general with its strong sense of neighbourhood, professional affiliation among the inhabitants and spatial elements of the circulation system (*Bunbast* and the *Maindânche)* of the neighbourhood.

Lessons

> Emphasising the values embodied in traditional cities does not automatically imply rejecting evolutionary forms of changes, nor does it mean that all traditional structures should be conserved at any cost. But it does suggest that certain essential structuring principles can be revived, adapted, and perpetuated to the advantage of contemporary societies. (Bianca 2000: 330)

The research identified the process of city formation, its overall structure and the course of the principal road as the backbone of the city. These findings can be seen as new and valuable knowledge to build a background for further research in particular relating to the city's conservation plan and its development. The application of these findings can be actualised when it is supposed to have some changes over the built form of the city's historic core. In designing the first crossroads of the city, one can realise how the lack of this background information alongside the main trend of modernisation (modern planning) led to destruction of the city's spatial configuration, especially how the bazaar complex as unifying element has been seriously damaged (Fig. 3.6, aerial photo).

Considering the structural elements of the neighbourhood and the principal road, some principles were identified in relation to the concept of centre originated from the notions of line and point. Where a number of socio-spatial factors involved, the concept of centre emerged in a linear form and where it was bound up with only one factor, the centre took a concentrated form. These principles can be recognised as indigenous rules embedded in the structure of the traditional core which can help designers to structure the centres of new urban areas or restructure the existing ones.

The concept of mound cities was conceptualised as the Kurdish style of urban planning and the fact that Kurdish culture has strong responsiveness to the Genius Loci, to the setting of the city as a whole and to the dwellings in a particular relationship to the landscape. This concept can be recognised as a symbolic value of Kurdish culture inherited from pre-Islamic times. From this point, it is the responsibility of urban designers to reconcile that concept with the contemporary urban architecture of the city. New developments of the city in the lowlands can possibly take advantage of this concept by placing the symbolic and memorial elements at the more elevated parts of the site. Furthermore, in laying out new suburbs, any changes of topography should capture the percieved relationship with the natural landscape. Here the value of landscape is such that the designer does not need to simulate the landscape in the courtyard. The value of landscape, in itself, invites the designer to work with not against the topography. This means structuring the form of terraced houses to enable the dwellings to be opened up to the natural landscape and the public domain by use of the element of a *Bar-haiwan*. This principle is also in line with a main notion of urban design which regards the ecological significance of place as vital for preserving the natural environment. More importantly, the value of *the Bar-haiwan* concept as a Buffer zone between private and public domains enables the residents not to have to restore to covering their windows, which in turn increases the residents' ability to be in contact with the public domain and encourages social activities.

The sceno-graphic approach to the street pattern of the neighbourhood, which exposed different forms of gateway to the passer-by in the circulation system of the quarter, can be considered as a rule for neighbourhood design in order to enhance the visual qualities. Concerning this rule, one principle would be to work with gateways location along the circulation system of the new neighbourhood, especially where the low-lands are targeted for the expansion of the city and the view of landscape is limited because of the setting. From this point, the designers would have to avoid long straight roads to enable the plan to create more opportunities for a variety of images of doorways to be espoused to the view of passers-by, thus, enhancing the experiential qualities of space. The concept of the *Maidanche*, as a buffer zone between private and public domains (Abu-Ghazzeh, 1999), within the cluster houses can be recognised as another structural principle which, on the one hand, can help to enhance the experiential qualities of space in the circulation system of the neighbourhood and, on the other hand, to strengthen the sense of neighbourhood and social interactions among the inhabitants. This concept forms "proximity (functional and physical)" as a prerequisite to encourage social interactions among the neighbours. (Kearns & Parkinson, 2001:104; Williams, 2005: 197). As studies show, "people communicate more through the Internet, and collaborate more effectively, when they are in closer proximity" (Kelly et al., 2012: 10). For this reason, "increasing proximity through design increases repeated passive contacts between residents, which [in turn] helps to form social relations" (Williams 2005: 197).

The self-reliant character of the administrative structure of the neighbourhood, arose mainly from the professional affiliation among the inhabitants and the sense of community it nurtured. It is a significant lesson that illustrate how those who experienced the place made decisions from bottom up. They sustained the society without recourse to external agencies. This means that the local communities in contemporary society can be empowered to nurture their own development plans and strategies from within and "overcome the division between subject and object" in today's urban planning (Bianca, 2000: 337). In line with this point, one principle would be to empower the local voice, especially in the neighbourhoods, by increasing the commitment to the area (Corcoran, 2002). This is consistent with the view of Akbar (1988), who believes that the shift of responsibility from residents to the state, which has no direct experience of place, has created the crisis in the contemporary built environment of Muslim cities. This means that, in the modern form of urban management in Iran, control of the environment is far less than in that of the traditional form. Hence, urban management must be shifted away from a state-centred perspective to a form of urban governance that has "its focus on the nature of the relationship between different actors in the urban area" and that mediates among them on the basis of conviction and of responsibility (Corubolo, 1998: 14, da Cruz et al., 2019). In such a bottom-up approach, the neighbourhoods would have a real opportunity to articulate their views about what matters to them concerning their place. Consequently, decisions and proposals could address local issues and priorities and "enable sensitivity and deployment of indigenous [or local] knowedge in planning"/design (Lane & McDonald, 2005: 710).

Conclusion Remark

> The book has ended but the story is ongoing
> It is impossible to reveal one's enthusiasm in a hundered books.[2]

Although the theme of this book was ended with some lessons for contemporary urban planning and design, based on the above words of Sa'di Shirazi, it is still a first step in the conceptualisation of the

[2] Sheikh Muslih-uddin Sa'di Shirazi: world famous Iranian thirteenth century poet.

nature of Kurdish cities and hopefully an opened window for further research in near future concerning other Kurdish cities in the region. Within the wider context, a comparative approach within Kurdish cities can highlight similarities and differences to form their main socio-spatial characteristics and underlying design principles. It is also hoped that the principles identified will reveal a broader picture of the cultural background for designers to mitigate the tension between tradition and modernity by creating a link to the context in new development plans as the city develops and prospers.

References

Abu-Ghazzeh, T. M. (1999). Housing layout, social interaction and the place of contact in Abu-nuseir, Jordan. *Journal of Environmental Psychology, 19*, 14–73.

Akbar, J. (1988). *Crisis in the built environment: The case of the Muslim city*. Concept Media

Alizadeh, H. (2007). Changes conceptions of women's public space in the Kurdish city. *Cities, 24*(6), 410–421. https://doi.org/10.1016/j.cities.2007.06.002

Bianca, S. (2000). *Urban form in the Arab world: Past and present*. Vdf.

Corcoran, M. P. (2002). Place attachment and community sentiment in marginalised neighbourhoods: A European case study. *Canadian Journal of Urban Research*, 47–67.

Corubolo, E. (1998). *Urban management and social Justice* (p. 21). Working Paper No. 92. London, Development Planning Unit, University College London. https://eprints.ucl.ac.uk/archive/00000033/01/wp92.pdf.

da Cruz, N. F., Rode, P., & McQuarrie, M. (2019). New urban governance: A review of current themes and future priorities. *Journal of Urban Affairs, 41*(1), 1–19.

Izady, M. R. (1992). *A concise handbook: The Kurds*. Crane Russak (Taylor & Francis).

Kearns, A., & Parkinson, M. (2001). The significance of neighbourhood. *Urban Studies, 38*(12): 2103–2110.

Kelly, J. F., Breadon, P., Davis, C., Hunter, A., Mares, P., Mullerworth, D., & Weidmann, B. (2012). *Social cities*. Grattan Institute.

Kostof, S. (1991). *The city shaped: Urban patterns and meanings through history.* Thames & Hudson.

Kostof, S. (1992). *The city assembled: The elements of urban form through history.* Thames & Hudson.

Kreyenbroek, P. G. (1996). Religion and religions in Kurdistan. In P. G. Kreyenbroek & C. Allison (Eds.), *Kurdish culture and identity* (pp. 85–110). Zed Books in association with The Centre of Near and Middle Eastern Studies.

Lane, M. B., & McDonald, G. (2005). Community-based environmental planning: Operational dilemmas, planning principles and possible remedies. *Journal of Environmental Planning and Management, 48*(5), 709–731.

Minorsky, V. (1943). Sanandaj. In M. T. Houtsma, A. J. Wensinck, H. A. R. Gibb, W. Heffening & E. Levi-Provencal (Eds.), *Encyclopaedia of Islam.* E. J. Brill and Leyden & Luzac & Co.

Morris, A. E. J. (1994). *History of urban form: Before the industrial revolutions.* Longman Scientific & Technical and Wiley.

Norberg-Schulz, C. (1980). *Genius Loci: Towards a phenomenology of architecture.* Rizzoli International Publications.

Rapoport, A. (1977). *Human aspects of urban form: Towards a man-environment approach to urban form and design.* Pergamon Press.

Vasilyeva, EI. (2000) The Kurdish city of Sanandaj: Ardelan's Capital. Paris, conference presentation.

Williams, E. (2000). Baths and bathing culture in the Middle East: The Hammam. In *Heilbrunn timeline of art history.* The Metropolitan Museum of Art, 2000–. http://www.metmuseum.org/toah/hd/bath/hd_bath.htm. (October, 2012).

References

Abu-Ghazzeh, T. M. (1994). Built form and religion: Underlying structure of Jeddah Al-Qademah. *Traditional Dwellings and Settlements Review, V*(II): 49–55.

Abu-Ghazzeh, T. M. (1999). Housing layout, social interaction and the place of contact in Abu-nuseir, Jordan. *Journal of Environmental Psychology, 19*, 14–73.

Abu-Lughod, J. (1983). Contemporary relevance of Islamic urban principles. In A. Germen (Ed.), *Islamic architecture and urbanism* (pp. 64–70). King Faisal University.

Ahmad, K., & Malcolm, P. (2001). Traditional settlements, user participation and 'no harm' principle toward a contemporary participatory process in low-income public housing in Islamic cities. In H. Turgut & P. Kellett (Eds.), *Traditional environments in a new millennium* (pp. 69–74). Istanbul Technology University.

Akbar, J. (1988). *Crisis in the built environment: The case of the Muslim city.* Concept Media

Alemi, M. (1991). Urban spaces as the scene for the Ceremonies and pastimes of the Safavid court. *Journal of the Islamic Environmental Design Research Centre, 11*(2), 98–107.

H. Alizadeh, *Building a traditional Kurdish City*,
https://doi.org/10.1007/978-981-16-3634-9

Alizadeh, H. (2004). The Kurdish city of Sanandaj. *Journal of Kurdish Studies, 5*, 103–126.

Alizadeh, H. (2007). Changes conceptions of women's public space in the Kurdish city. *Cities, 24*(6), 410–421. https://doi.org/10.1016/j.cities.2007.06.002

Alizadeh, H., & Irandoost, K. (2017). Investigating the process of traditional design principles formation in the Iranian-Kurdish urban quarters. *Urban Design International, 22*(3), 197–218.

Alizadeh, H., & Habibi, K. (2008). Structural elements and the built environment. *International Journal of Environmental Research, 2*(2).

Alizadeh, H., Bahrami, B., & Farahani, L. M. (2019). Geophysical setting and urban form of a traditional city: Examination of the Kurdish city of Sanandaj, Iran. *Proceedings of the Institution of Civil Engineers-Urban Design and Planning, 172*(6), 237–247.

Al-Kodmany, K. (2000). Women's visual privacy in traditional and modern neighbourhoods in Damascus. *Architecture and Planning Research, 17*(4), 283–303.

Allison, C. (1996). Old and new oral tradition in Badinan. In P. G. Kreyenbroek & C. Allison (Eds.), *Kurdish culture and identity* (pp. 29–47). Zed Books in association with The Centre of Near and Middle Eastern Studies.

Alten, I. (1958). *Sanandaj, a town plan.* United State Operation Mission to Iran.

Allison, C. (2001). *The Yezidi oral tradition in Iraqi Kurdistan.* Routledge.

Amin, A. (2008). Collective culture and urban public space. *City, 12*(1), 5–24.

Ardalan, N. (1980a). The visual language of symbolic form: A preliminary study of mosque architecture. In J. G. Katz (Ed.), *Architecture as symbol and self-identity*. Aga Khan Award for Architecture.

Ardalan, N. (1980b). Places of public gathering. In L. Safran (Ed.), *Places of public gathering in Islam.* Aga Khan Award for Architecture.

Ardalan, S. (2008). Kurdish tribe of Ardalan: Iran and Ottoman Empire conflicts (M. Ardalan, Trans.). Iran History Publication.

Ardalan, N., & Bakhtiar, L. (1973). *The sense of unity: The sufi tradition in Persian architecture*. University of Chicago Press.

Arefian, F. F., & Moeini, S. H. I. (2016). *Urban change in Iran.* Springer.

Ayazi, B. (1992). *Ainah-yi Sanandaj*. Ayazi, Burhan.

Babani, A., (Ed.). (1998). *Saerull-Akrad dar Tarikh va Guqkraphia-E- Kurdistan* (An overview of history and geography of Kurdistan). Awrang Press.

Bahrami, B. (2001). *Tarahei Maskan dar Mahalla-ye Sartapula-ye Sanandaj* (Housing design in the neighbourhood of Sartapulah, Sanandaj). (Master), Art University of Isfahan, Isfahan.

Bakhtiar, A. A. (2011). Reminiscences of the Maidan-e Shah. In C. P. Mitchell (Ed.), *New perspectives on Safavid Iran: Empire and society*. Routledge.

Barth, F. (1953). Principles of social organization in Southern Kurdistan. Universitetets Etnografiske Museum. *Bulletin, 7*, 26.

Baskaya, A., & Symes, M. (1994). *A study of a house type with an open space. People, place and development*. CARDO.

Bayat, A. (1988). *Kuliate Guqkrafia- E- Tabeeai va Tarikhi Iran* (A historical and physical geography of Iran). Sepehr Press.

Benton-Short, L., & Short, J. R. (2013). *Cities and nature*. Routledge.

Bianca, S. (2000). *Urban form in the Arab world: Past and present*. Vdf.

Boardman, P. (1978). *The worlds of Patrick Geddes: Biologist, town planner, re-educator, peace-warrior*. Routledge.

Bonine, M. E. (1979). The morphogenesis of Iranian cities. *Annals of the Association of American Geographers, 69*(2), 208–224.

Boyce, M. (2001). *Zoroastrians: Their religious beliefs and practices*. Routledge.

Browne, K. (1976). Making of the city: A special issue on Isfahan. *Architectural Review, 951*(v.CLIX), 259–291.

Bruinessen, M. V. (1978). *Agha, Shaikh and state: On the social and political organization of Kurdistan*. Rijksuniversiteit.

Bruinessen, M. V. (2001). From Adela Khanum to Leyla Zana: Women as political leaders in Kurdish history. In S. Mojab (Ed.), *Women of a non-state nation: The Kurds* (pp. 95–112). Mazda Publishers.

Clark, J. I., & Clark, B. D. (1969). *Kermanshah: An Iranian provincial city*. University of Durham.

Conner, P. (1987). On the bath: Western experience of the Hammam. *Culture, Theory and Critique, 31*(1), 34–42.

Corcoran, M. P. (2002). Place attachment and community sentiment in marginalised neighbourhoods: A European case study. *Canadian Journal of Urban Research*, 47–67.

Corubolo, E. (1998). *Urban management and social Justice* (p. 21). Working Paper No. 92. London, Development Planning Unit, University College London. https://eprints.ucl.ac.uk/archive/00000033/01/wp92.pdf.

Cresswell, T. (2014). *Place: An introduction*. Wiley.

da Cruz, N. F., Rode, P., & McQuarrie, M. (2019). New urban governance: A review of current themes and future priorities. *Journal of Urban Affairs, 41*(1), 1–19.

Davis, H. (1997). Location and site. In P. Oliver (Ed.), *Encyclopedia of vernacular architecture of the world* (pp. 140–143). Cambridge University Press.

Dultshahi, A. (1984). *Guqkrapgiai Qkarbi Iran ya Kuhhai Nashnahkti Evista.* Naghshi Jahan.

Edwards, B., Sibley, M., Land, P., & Hakmi, M. (Eds.). (2006). *Courtyard housing: Past, present and future*. Taylor & Francis.

English, P. W. (1966). *City and village in Iran: Settlement and economy in the Kirman Basin.* University of Wisconsin Press.

Ergecgil, S. C. (2001). Modern culture and its perception of tradition, history and historical sites. In H. Turgut & P. Kellett (Eds.), *Traditional environments in a new millennium* (pp. 75–78). Istanbul Technology University.

Falamaki, M. (1987). Sheklgiry Shahrhaie Iran (Iranian cities formation). In M. Y. Kiani (Ed.), *Nazari Ijmali bih Shahr-nishini va Shahrsazi dar Iran* (A general study on urbanisation and urban planning in Iran) (pp. 1–21). Farhang va Ershad-e Eslami.

Falamaki, M. (1995). *Baz-Zindasazie Banaha va Shahrha-ye Tarikhi* (Renovation of buildings and historical cities). Tehran University Press.

Far-Afza. (1991). *Tarh-e Tawseah va Omran va Hawzaieh Nufuz va Tafseli-e Shar-e Sanandaj* (Sanandaj's comprehensive and master plan). Sazman-e Maskan va Shahrsazi.

Galletti, M. (2001). Western images of women's role in Kurdish society. In S. Mojab (Ed.), *Women of a non-state nation: The Kurd* (pp. 209–225). Mazda Publishers Inc.

Garthwaite, G. R. (1981). Khans and Kings: The dialectics of power in Bakhtiyari history. In N. R. Keddie (Ed.), *Modern Iran.* State University of New York Press.

Gaube, H. (1979). *Iranian cities.* New York University Press.

Gehl, J. (2011). *Life between buildings: Using public space*. Island Press.

Ghandchi, S. (1999). *Kurds and formation of central government in Iran.* http://www.ghandchi.com/02-Kurds/KurdsEnglish.htm. 2003.

Gieryn, T. F. (2000). A space for place in sociology. *Annual Review of Sociology, 26*, 463–496.

Grabar, O. (1984). Palaces, citadels and fortifications. In G. Michell (Ed.), *Architecture of the Islamic world: Its history and social meaning* (pp. 65–79). Thames & Hudson.

Grabolle-Çeliker, A. (2019). Kurdish women. In M. M. Gunter (Ed.), *Routledge handbook on the Kurds* (pp. 239–256). Routledge.

Grandmaison, C. L. C. (1977). Spatial organisation, tribal groupings and kinship in Ibra. *Journal of Oman Studies, 3*(2), 95–106.

Guest, J. S. (1993). *Survival among the Kurds: A history of the Yezidis*. London: Routledge.

Gunter, M. M. (2018). *Historical dictionary of the Kurds* (Vol. 8). Rowman & Littlefield.

Hambly, G. R. G. (1968). Agha Muhammad Khan and the establishment of the Qajar dynasty. In W. B. Fisher (Ed.), *The Cambridge history of Iran* (Vol. 7, pp. 104–174). Cambridge University Press.

Habibi, S. M. (1996). *Az Shar Ta Shahr* (Analytical review of the city concept and its physical image in the course of time). University of Tehran.

Hakim, B. S. (1986). *Arabic-Islamic cities: Building and planning principles*. KPI.

Hall, T., & Barrett, H. (2012). *Urban geography*. Routledge.

Hambly, G. R. G. (1991). The traditional Iranian city in the Qajar period. In P. Avery, G. Hambly, & C. Melville (Eds.), *The Cambridge history of Iran: From Nadir Shah to the Islamic republic* (Vol. 7, pp. 542–589). Cambridge University Press.

Hansen, H. H. (1960). *Daughters of Allah*. Purnell.

Haqshenas, M. (1999). *Sharakam Sennah* (Sanandaj; my land). Zhyar.

Hassanpour, A. (1996). The creation of Kurdish media culture. In P. G. Kreyenbroek & C. Allison (Eds.), *Kurdish culture and identity* (pp. 45–84). Zed Books in association with The Centre of Near and Middle Eastern Studies.

Hardi, A. (2004). The secret of solitude. *The International Journal of Kurdish Studies, 18*(1–2), 192.

Hay, W. R. (1921). *Two years in Kurdistan: Experiences of a political officer 1918–1920*. Sidgwick & Jackson Ltd.

Holland, C., Clark, A., Katz, J., & Peace, S. (2007). *Social interactions in urban public places*. Policy Press.

Hossain, M. Z. (2015). Water: The most precious resource of our life. *Global Journal of Advanced Research, 2*(9), 1–11.

Izady, M. R. (1992). *A concise handbook: The Kurds*. Crane Russak (Taylor & Francis).

Jiven, G., & Larkham, P. J. (2003). Sense of place, authenticity and character: A commentary. *Journal of Urban Design, 8*(1), 67–81.

Kamandi, A. (2014). *Pas Koochahaee Sanandaj* (Sanandaj's alleys). College Scientific Publication (in Persian).

Katouzian, S. (1996). Tehran, capital city: 1786–1997 the re-invitation of a metropolis. *Journal of the Islamic Environmental Design Research Centre, 15*, 34–44 (In L. Micara & M. Alemi (Eds.), Como, Italy).
Kearns, A., & Parkinson, M. (2001). The significance of neighbourhood. *Urban Studies, 38*(12): 2103–2110.
Kellett, P. (2003). Constructing informal places. In S. Menin (Ed.), *Constructing place: Mind and matter* (pp. 87–98). Routledge.
Kelly, J. F., Breadon, P., Davis, C., Hunter, A., Mares, P., Mullerworth, D., & Weidmann, B. (2012). *Social cities*. Grattan Institute.
Khalid, A. (1991). The residential quarter in Bukhara before the revolution (The Work of O. A. Sukhareva). *Middle East Studies Association Bulletin, 25*(1), 15–24. https://doi.org/10.1017/S0026318400023622.
Khatib-Chahidi, J. (1981). Sexual prohibitions, shared space and fictive marriage in Shi'ite Iran. In S. Ardener (Ed.), *Women and space* (pp. 112–135). Croom Helm.
Khidir, O. M. O. (1998). *Bridging between the contemporary and the vernacular architecture* (p. 85) (Electronical Dissertation Architecture). Virginia Polytechnic Institute and State University.
Kheirabadi, M. (1991). *Iranian cities: Formation and development*. University of Texas Press.
Kheirabadi, M. (2004). *Religions of the world Islam*. Chelsea House Publishers.
Kiani, M. Y., (Ed.). (1986). *Shahr-ha-yi Iran* (Iranian cities). Jahad-e Danishgahi.
Kiani, M. Y., (Ed.). (1987). *Nazari Ijmali bih Shahr-nishini va Shahrsazi dar Iran* (A general study on urbanisation and urban planning in Iran). Farhang va Ershad-e Eslami.
Kinneir, S. J. M. (1813). [Republished 1973] *A Geographical memoir of the Pesian Empire: Journey through Asia Minor, Armenia, and Koordistan in the years 1813 and 1814*. Arno Press.
Kirmanj, S. (2014). Kurdish history textbooks: Building a nation-state within a nation-state. *The Middle East Journal, 68*(3), 367–384.
Kite, S. (2002). The poetics of Oman's traditional architecture: Towards an aesthetic interpretation. *Journal of Oman Studies, 12*, 133–155.
Kite, S. (2008). Mountain and paradise: Wall and court, space and transition, in Iran and Oman. In H. Alizadeh (Ed.), *International Conference on Zagros Traditional Settlements* (pp. 25–34). Sanandaj, University of Kurdistan.
Kostof, S. (1991). *The city shaped: Urban patterns and meanings through history*. Thames & Hudson.

Kostof, S. (1992). *The city assembled: The elements of urban form through history.* Thames & Hudson.

Krehl, A., Siedentop, S., Taubenböck, H., & Wurm, M. (2016). A comprehensive view on urban spatial structure: Urban density patterns of German city regions. *ISPRS International Journal of Geo-Information, 5*(6), 76.

Kreyenbroek, P. G. (1996). Religion and religions in Kurdistan. In P. G. Kreyenbroek & C. Allison (Eds.), *Kurdish culture and identity* (pp. 85–110). Zed Books in association with The Centre of Near and Middle Eastern Studies.

Kurdistani, A. V. (2002). *Hadigha-E-Nasriah dar Guqkrafia VA Tarikh-e Kurdistan* (History and geography of Kurdistan) (Persian). Tavakouli.

Lambton, A. S. (1968). Land tenure and revenue administration in the ninteenth century. In W. B. Fisher (Ed.), *The Cambridge history of Iran* (Vol. 7, pp. 459–505). Cambridge University Press.

Lambton, A. S. (1991). *Landlord and peasant in Persia: A study of land tenure and land revenue administration.* Martin Press.

Lambton, A. S. (2011). Administration and social organisation. In E. Yarshater (Ed.), *Encyclopaedia Iranica: Cities.* Columbia University and Bibliotheca Persia.

Lane, M. B., & McDonald, G. (2005). Community-based environmental planning: Operational dilemmas, planning principles and possible remedies. *Journal of Environmental Planning and Management, 48*(5), 709–731.

Lapidus, I. M. I. M. (1967). *Muslim cities in the later middle ages.* Harvard University Press.

Lawson, B. (2001). *The language of space.* Architectural Press.

Lawson, B. (2007). *The language of space.* Architectural Press.

Leatherbarrow, D. (1993). *The roots of architectural invention.* Cambridge University Press.

Le Strange, G. (1905). *The lands of the Eastern caliphate: Mesopotamia, Persia, and central Asia, from the Moslem conquest to the time of Timur.* Cambridge University Press.

Levkowitz, J. (2018). Palaces for the people: Scrutinizing social infrastructure in Sulaimani. *The Journal of Intersectionality, 2*(2), 24–32.

Levine, L. D. (1974). Geographical studies in the Neo-Assyrian Zagros - I I. *Persian Studies, 12*, 19–123.

Lockhart, L. (1939). *Famous cities of Iran.* W. Pearce & Co.

Lombard, M. (2014). Constructing ordinary places: Place-making in urban informal settlements in Mexico. *Progress in Planning, 94*, 1–53.

Kiani, M. Y., (Ed.). (1986). *Shahr-ha-yi Iran* (Iranian cities). Jahad-e Danishgahi.
Madanipour, A. (1998). *Tehran: The making of a metropolis*. Wiley.
Madanipour, A. R. (2003). *Public and private spaces of the city*. Routledge.
Madanipour, A. (2013). Public spaces of European cities. *NA*, *18*(1).
McDowall, D. (1996). *The Kurds*. Minority Rights Group.
McDowall, D. (2001). *A modern history of the Kurds/translated by Abrahim unis*. Paniz Press.
Mardukh, M. (1972). *Tarikh-e- Kurd VA Kurdistan* (The history of Kurds and Kurdistan). Qkariqhi (in Persian).
Mardukh, M. (2000). *Tarikh-e- Mardukh* (The history of Kurds and Kurdistan). Karang.
Marjan. (1977). *Tarh-e Jamia Shar-e Sanandaj* (Sanandaj's master plan). Kurdistan Urban Development Organization.
Mazumdar, S., & Mazumdar, S. (2001). Rethinking public and private space: Religion and women in Muslim society. *Journal of Architectural and Planning Research, 18*(4), 302–324.
Mazumdar, S., & Mazumdar, S. (2004). Religion and place attachment: A study of sacred places. *Journal of Environmental Psychology, 24*(3), 385–397.
Mehriar, M., Fatholayouf, S., Tehrani, F. F., & Qaderi, B. (1998). *Asnad-e-Tasveri shahrhaei Irani* (The old maps of Iranian cities). Shahid Beheshti University Press.
Memarian, G., & Brown, F. E. (2003). Climate culture and religion: Aspects of the traditional courtyard house in Iran. *Architecture and Planning Research, 20*(3), 180–198.
Minorsky, V. (1943). Sanandaj. In M. T. Houtsma, A. J. Wensinck, H. A. R. Gibb, W. Heffening & E. Levi-Provencal (Eds.), *Encyclopaedia of Islam*. E. J. Brill and Leyden & Luzac & Co.
Mirawdeli, K. (2003). Claudius Rich's Kurdistan in 1820. *Kurdistanica* (The Encyclopaedia of Kurdistan), History: Related articles. Nottingham. https://www.kurdistanica.com/index.html.
Mizanian, N. (2017). Mastoureh Ardalan, a rarity of her time, the first woman historian. *Tehran Times*. Mehr News Agency.
Mojab, S. (2003). Kurdish women. In S. Joseph (Ed.), *Encyclopedia of women and Islamic cultures: Family, law and politics* (Vol. 2, pp. 358–366). Brill.
Monteguin, F.-A. D. (1983). The essence of urban existence in the world of Islam. In A. Germen (Ed.), *Islamic architecture and urbanism* (pp. 43–63). King Faisal University.

Moradi, A. M., & Amirkabirian, A. (2001). *Muarafei Teadadi az Abniea-e Sunat-e Iran va Tahlili bar Vizhagihaie Fazaei Anha* (Introducing a number of traditional buildings in Iranian architecture and analysis of their spatial characteristics). Pazhuheshgah.

Morris, A. E. J. (1994). *History of urban form: Before the industrial revolutions*. Longman Scientific & Technical and Wiley.

Moughtin, C. (2003). *Urban design: Street and square*. Architectural Press.

Najimi, A. W. (1988). *Herat, the Islamic city: A study in urban conservation*. Curzon Press LTD.

Nikitine, V. (1983). Kurd and Kurdistan. Translated by M. Qazi, Tehran, Nilufar.

Nikitine, B. (1956). *Les Kurdes. Etude sociologique et historique*. Imprimerie nationale, Bibliothèque C. Klincksieck.

Norberg-Schulz, C. (1971). *Existence, space & architecture*. Studio Vista.

Norberg-Schulz, C. (1980). *Genius Loci: Towards a phenomenology of architecture*. Rizzoli International Publications.

Oldenburg, R. (1989). *The great good place: Cafés, coffee shops, community centers, beauty parlors, general stores, bars, hangouts, and how they get you through the day*. Paragon House Publishers.

Oliver, P. (1983). Binarism in an Islamic city: Isfahan as an example of geometry and duality. In *Islamic Architecture and Urbanism. A. Germen (ed). Dammam,* (pp. 121–146). King Faisal University.

Oliver, P., Davis, I., & Bentley, I. (1981). *Dunroamin: Thesuburban semi and its enemies*. Barrie & Jenkins.

Ozaslan, N. (1996). *Historic urban fabric: Source of inspiration for contemporary city form* (p. 373) (Dept. of Architecture. Unpublished Ph.D). University of York.

Pars, B. (2004). *Tarhe Samandah-ye Site-e Firdaus-ye Sanandaj* (Revitalisation of the Sanandaj Firdausi's site). Kurdistan Housing and Urban Planning Organization.

Peernia, M. K. (1995). *Ashnaei ba Meamari-e Islami: Sakhtimanhai Darwn-Shahri va Burwn-Shahri* (Recognising the Islamic architecture). University of Elmuo-Sanat.

Perry, J. (1968). The Zand dynasty. In W. B. Fisher (Ed.), *The Cambridge history of Iran* (Vol. 7, pp. 63–103). Cambridge University Press.

Pourjafar, M., Amini, M., Varzaneh, E. H., & Mahdavinejad, M. (2014). Role of bazaars as a unifying factor in traditional cities of Iran: The Isfahan bazaar. *Frontiers of Architectural Research, 3*(1), 10–19.

Rainer, R. (1977). *Traditional building in Iran*. Akademische Druck- u.

Rangavary. (1999). City's memories/Interviewer: M. Haghshenas, S. Marufi, & S. Munaghi. (Vol 1). Sanandaj.

Rapoport, A. (1977). *Human aspects of urban form: Towards a man-environment approach to urban form and design*. Pergamon Press.

Rapoport, A. (1989). On the attributes of tradition. In J.-P. Bourdier & N. Alsayyad (Eds.), *Dwellings, settlements and tradition: Cross-cultural perspectives* (pp. 77–105). University Press of America.

Rich, C. J. (1836). *Narrative of a residence in Koordistan, and on the site of ancient Nineveh: With journal of a voyage down the Tigris to Bagdad and an account of a visit to Shirauz and Persepolis*. J. Duncan.

Roemer, H. R. (1968). The Safavid period. In W. B. Fisher (Ed.), *The Cambridge history of Iran* (Vol. 6, pp. 189–350). Cambridge University Press.

Rossi, A. (1982). *Architettura della città*. MIT Press.

Ross, N., & Mohammadpur, A. (2018). Imagined or real: The intersection of tribalism and nationalism in the Kurdish Regional Government (KRG). *British Journal of Middle Eastern Studies, 45*(2), 194–211.

Safrastian, A. (1948). *Kurds and Kurdistan*. The Harvill Press LTD.

Saleh, M. A. E. (1995). Al-Alkhalaf: The urban history of a traditional settlement in Southwestern Saudi Arabia. *Habitat International, 19*(1): 29–52.

Saleh, M. A. E. (2001). Planning and designing for defense, security and saftey in Saudi Arabian residential environments. *Architecture and Planning Research, 18*(1), 39–58.

Sanandaji, M. S. (1996). *Tuhfa-ye Nasiriye dar Tarikh va Jughrafiya-yi Kurdistan* (Kurdistan history and geography). Amir-Kabir.

Sarmento, J., & Kazemi, Z. (2014). Hammams and the contemporary city: The case of Isfahan, Iran. *International Journal of Heritage Studies, 20*(2), 138–156.

Saravand. (1992). *Tarhe Amadasazei 150 Hectare Gharadyan* (Land development plan). Sâzmân-e Maskan va Shahrsazi Kurdistan.

Setiadi H. (2015). Islam and urbanism in Indonesia: The mosque as urban identity in Javanese cities. In S. Brunn (Eds), *The changing world religion map* (pp: 2415–2416). Springer. https://doi.org/10.1007/978-94-017-9376-6_127.

Shah, R. (2015). *Making sense of disability: The role of home and its physical environment* (p. 261) (Doctor of Philosophy). School of Design. Creative Industries Faculty, Queensland University of Technology.

Shapur Shahbazi, A., & Edmund Bosworth, C. (1990). Capital cities. *Encyclopædia Iranica*. Online edition. Vol. IV, Fasc. 7

Sharafkandi, A. (2002). *Mejwi Ardalan* (The history of Ardalan). Taze Nghah press.

Simmel, G. (2007). The metaphysics of death. *Theory, Culture & Society, 24*(7–8), 72–77.

Soltani, A. (2006). The Sharafnama of Bitlisi manuscript copies, translations and appendixes. *The International Journal of Kurdish Studies, 20*(1–2), 209.

Southall, A. (1998). *The city in time and space*. Cambridge University Press.

Spirn, A. W. (1988). The poetics of city and nature: Towards a new aesthetic for urban design. *Landscape Journal, 7*(2), 108–126.

Stewart, D. J. (2001). Middle east urban studies: Identity and meaning. *Urban Geography, 22*(2), 175–181.

Sultanzadeh, H. (1983). *Ravand-e Sheklgir-ye Shar va Marakiz Mazhabi dar Iran* (The process of City and Religious Centres Formation in Iran). Agah Press.

Sultanzadeh, H. (1986). *Moghadamai Bar Tariekhi Shahr va Shahrnishini dar Iran* (An introduction into the history of the city and urbanisation in Iran). Amir Kabir Press.

Sultanzadeh, H. (2001). *Iranian bazaars*. Cultural Research Bureau.

Tallan, C., & Taitz, E. (2009). *Learned women in traditional Jewish society in Jewish women: A comprehensive historical encyclopedia*. https://jwa.org/encyclopedia/article/learned-women-in-traditional-jewish-society. Online: Jewish women's Archive.

Tapper, N. (1990). Ziyaret: Gender, movement, and exchange in a Turkish community. In *Muslim travellers: Pilgrimage, migration, and the religious imagination* (pp. 236–255).

Tjahjono, G. (1989). Centre and duality in the Javanese Dwelling. In J.-P. Bourdier & N. Alsayyad (Eds.), *Dwellings, settlements and tradition: Cross-cultural perspectives* (pp. 213–236). University Press of America.

Tavassoli, M. (1982). *Sakht-i Shahr va Miamari dar Iqlim-i Garam va Khushk-i Iran* (Urban structure and architecture in the Hot-arid Zone of Iran). Urban Planning and Architecture Research Centre of Iran.

Tavassoli, M., & Bonyadi, N. (1991). *Urban Space Design1*. Urban Planning and Architecture Research Centre of Iran.

Tavassoli, M., & Bonyadi, N. (1992). *Urban space design 2*. Urban Planning and Architectural Research Centre of Iran.

Tuan, Y.-F. (1977). *Space and place: The perspective of experience*. University of Minnesota Press.

Vaceq Abbasi, A., & Abdi, M. R. (2017). TUS, the golden footed. *The Journal of Epic Literature* (PAZHUHESH-NAMEH FARHANG-O-ADAB), *13*(1 [23]), 89–113.

Vasilyeva, EI. (2000) The Kurdish city of Sanandaj: Ardelan's Capital. Paris, conference presentation.

Warren, J., & Fethi, I. (1982). *Traditional houses in Baghdad*. Coach.

Williams, E. (2000). Baths and bathing culture in the Middle East: The Hammam. In *Heilbrunn timeline of art history*. The Metropolitan Museum of Art, 2000–. http://www.metmuseum.org/toah/hd/bath/hd_bath.htm (October, 2012).

Zylstra, G. D. (2017). Productions of space, productions of power: Studying space, urban design, and social relations. *Journal of Urban History, 43*(3), 562–569.

Index

H. Alizadeh, *Building a traditional Kurdish City*,
https://doi.org/10.1007/978-981-16-3634-9

The manufacturer's authorised representative in the EU is Springer Nature Customer Service Centre GmbH, Europaplatz 3, 69115 Heidelberg, Germany. If you have any concerns regarding our products, please contact ProductSafety@springernature.com

Printed and bound by CPI Group (UK) Ltd, Croydon, CR0 4YY

30/09/2026

02225455-0006